Springer Theses

Recognizing Outstanding Ph.D. Research

For further volumes:
http://www.springer.com/series/8790

Aims and Scope

The series "Springer Theses" brings together a selection of the very best Ph.D. theses from around the world and across the physical sciences. Nominated and endorsed by two recognized specialists, each published volume has been selected for its scientific excellence and the high impact of its contents for the pertinent field of research. For greater accessibility to non-specialists, the published versions include an extended introduction, as well as a foreword by the student's supervisor explaining the special relevance of the work for the field. As a whole, the series will provide a valuable resource both for newcomers to the research fields described, and for other scientists seeking detailed background information on special questions. Finally, it provides an accredited documentation of the valuable contributions made by today's younger generation of scientists.

Theses are accepted into the series by invited nomination only and must fulfill all of the following criteria

- They must be written in good English.
- The topic should fall within the confines of Chemistry, Physics, Earth Sciences and related interdisciplinary fields such as Materials, Nanoscience, Chemical Engineering, Complex Systems and Biophysics.
- The work reported in the thesis must represent a significant scientific advance.
- If the thesis includes previously published material, permission to reproduce this must be gained from the respective copyright holder.
- They must have been examined and passed during the 12 months prior to nomination.
- Each thesis should include a foreword by the supervisor outlining the significance of its content.
- The theses should have a clearly defined structure including an introduction accessible to scientists not expert in that particular field.

Amit Finkler

Scanning SQUID Microscope for Studying Vortex Matter in Type-II Superconductors

Doctoral Thesis accepted by
the Weizmann Institute of Science,
Rehovot, Israel

Author
Dr. Amit Finkler
Weizmann Institute of Science
Rehovot
Israel

Supervisor
Prof. Eli Zeldov
Weizmann Institute of Science
Rehovot
Israel

ISSN 2190-5053
ISBN 978-3-642-29392-4
DOI 10.1007/978-3-642-29393-1
Springer Heidelberg New York Dordrecht London

ISSN 2190-5061 (electronic)
ISBN 978-3-642-29393-1 (eBook)

Library of Congress Control Number: 2012937647

© Springer-Verlag Berlin Heidelberg 2012
This work is subject to copyright. All rights are reserved by the Publisher, whether the whole or part of the material is concerned, specifically the rights of translation, reprinting, reuse of illustrations, recitation, broadcasting, reproduction on microfilms or in any other physical way, and transmission or information storage and retrieval, electronic adaptation, computer software, or by similar or dissimilar methodology now known or hereafter developed. Exempted from this legal reservation are brief excerpts in connection with reviews or scholarly analysis or material supplied specifically for the purpose of being entered and executed on a computer system, for exclusive use by the purchaser of the work. Duplication of this publication or parts thereof is permitted only under the provisions of the Copyright Law of the Publisher's location, in its current version, and permission for use must always be obtained from Springer. Permissions for use may be obtained through RightsLink at the Copyright Clearance Center. Violations are liable to prosecution under the respective Copyright Law.
The use of general descriptive names, registered names, trademarks, service marks, etc. in this publication does not imply, even in the absence of a specific statement, that such names are exempt from the relevant protective laws and regulations and therefore free for general use.
While the advice and information in this book are believed to be true and accurate at the date of publication, neither the authors nor the editors nor the publisher can accept any legal responsibility for any errors or omissions that may be made. The publisher makes no warranty, express or implied, with respect to the material contained herein.

Printed on acid-free paper

Springer is part of Springer Science+Business Media (www.springer.com)

List of Publications

1. Finkler, A., Segev, Y., Myasoedov, Y., Rappaport, M. L., Neeman, L., Vasyukov, D., Zeldov, E., Huber, M. E., Martin, J., and Yacoby, A. Self-aligned nanoscale SQUID on a tip. *Nano Letters* **10**, 1046 (2010).
2. Finkler, A., Vasyukov, D., Segev, Y., Neeman, L., Anahory, Y., Myasoedov, Y., Rappaport, M. L., Huber, M. E., Martin, J., Yacoby, A and Zeldov, E. Nano-sized SQUID-on-tip for scanning probe microscopy Accepted for publication in *Journal of Physics*: *Conference Series*.

Supervisor's Foreword

Study of magnetic behavior on nanoscale is of major scientific interest due to a wide range of fundamental quantum mechanical phenomena including quantum computation, spintronics, topological states of matter, and vortex dynamics in superconductors. There is a variety of techniques that provide macroscopic or microscopic, but sample-averaged, information on the magnetic state of the system, yet our present ability to determine the local magnetic field on the nanoscale is very limited. One of the major obstacles in study of nanomagnetism is the absence of readily accessible experimental methods for imaging the local magnetic fields and investigating their site-dependent dynamics on nanometer scale. Superconducting quantum interference devices (SQUID) have the highest field sensitivity, which is due in part to the large effective area of conventional SQUID probes. In recent years there is a growing interest in nanoSQUIDs and in scanning SQUID microscopy in which the field sensitivity is compromised for the benefit of spatial resolution and sensitivity to magnetic dipoles. Since the magnetic field of a dipole has $1/r^3$ dependence on the distance r, nanoSQUIDs are anticipated to become very sensitive detectors of magnetic moments with the ultimate goal of sensitivity of a single electron spin μ_B. Since the magnetic field of a nanostructure decays rapidly with distance, the achievable spatial resolution is determined not only by the size of the probe, but also to a large extent by the proximity of the probe to the sample. In order to achieve nanometer scale resolution the probe has to be able to approach and scan the sample within few nm from the surface. Such close proximity cannot be achieved with scanning probes based on planar lithographic technology.

The thesis of Amit Finkler presents a new scanning magnetic imaging method that provides quantitative, high-sensitivity mapping of static and dynamic magnetic fields on nanometer scale and is expected to achieve a combined performance that is significantly beyond the state of the art. The key element is the development of a unique method for fabrication of the smallest SQUID that resides on the apex of a very sharp tip. The fabrication process is based on a self-aligned deposition of a superconducting film onto a quartz tube pulled to a sharp pipette with aperture diameter as small as 100 nm. The resulting SQUID-on-tip made of aluminum has a flux sensitivity of 2×10^{-6} $\Phi_0/\text{Hz}^{1/2}$, magnetic field sensitivity of 10^{-7} $\text{T/Hz}^{1/2}$, bandwidth of about 1 MHz, and can operate at fields as high as 0.5 T. Moreover, due to its small size, this SQUID-on-tip has an outstanding spin sensitivity of 65 $\mu_B/\text{Hz}^{1/2}$. By gluing the SQUID-on-tip to a quartz tuning fork and monitoring its resonance frequency a sensitive feedback mechanism was developed which allows scanning of the tip at a constant distance of a few nm from the surface of the sample. Based on these achievements, the thesis describes the design and construction of a scanning SQUID microscope that provides simultaneous high resolution imaging of the sample topography and the local magnetic field. The potential of the microscope, which operates at a temperature of 300 mK over a wide range of magnetic fields, is demonstrated by the study of magnetic response of superconductors including imaging of the vortex lattice and the self-induced magnetic field generated by transport currents.

The novel SQUID-on-tip technique has prospect for significant further improvement in sensitivity and spatial resolution paving the way to a powerful tool for magnetic imaging on the nanoscale and single-spin-resolved scanning probe microscopy.

March 2012   Eli Zeldov

Acknowledgments

It is with pleasure and sincere thanks that I acknowledge the help, support, and encouragement I have received while working on this thesis.

First of all, Eli Zeldov has been a wonderful advisor. His acute physical intuition, vast knowledge and experience, and the ability to recall every missing detail in the experiment made a profound impact on me and helped in realizing this work from a small sketch on the whiteboard to a complete, functional system.

Yuri Myasoedov and Michael Rappaport contributed to the success of this work in more ways than I can remember. Their constant involvement and advice helped to speed up processes and proceedings that would have otherwise taken much longer to complete.

Much of the work in this thesis was done in collaboration with Yehonathan Segev. Working with Yehonathan has been truly fruitful and gainful, precisely because he and I are so different in our backgrounds and way of thinking. I am certain that this work would not have reached its present state without Yehonathan's participation.

One of the most crucial components of the system was fabricated by Martin E. Huber. His cryogenic SQUID array improved the signal-to-noise ratio tremendously. Martin has been exceptionally helpful in many of the system's most critical measurements. Our results would not have been as outstanding as they have had it not been for his joyful collaboration.

I am also especially grateful to the following people for their many contributions to my work and education: Nati Aharon, Jens Martin, Amir Yacoby, Gilad Barak, Hadar Steinberg, Sandra Foletti, and Michele Zaffalon.

Special gratitude is given to the Gas Liquefaction Center, the Physics Instrumentation workshop, the Electronics and Data Acquisition workshop, and to the members of the department's Secretariat.

My time in Eli's group has been enjoyable; I am grateful for being a part of this group, discussing and collaborating with past and present members: Nurit Avraham, Tal Verdene, Sarah Goldberg, Beena Kalisky, Haim Beidenkopf, Denis Vasyukov, Lior Neeman, Lior Embon, Ella Lachman, Yonathan Anahory, Jonathan Reiner, and Jonathan Drori.

Finally, I would like to thank my family for their support and love.

Contents

1 **Introduction** 1
   1.1 Scientific Background 1
      1.1.1 Vortices in Type-II Superconductors 1
      1.1.2 Superconducting Quantum Interference
            Device (SQUID) 7
      1.1.3 Tuning Fork Microscopy 9
   1.2 Open Questions 12
   1.3 Goal 14
   References 14

2 **Methods** 17
   2.1 SQUID-on-Tip Fabrication 17
   2.2 Tuning Fork Assembly 18
      2.2.1 Preparation 19
      2.2.2 Gluing a Tip and Its Effect on Resonance 20
      2.2.3 Mechanical Versus Electrical Excitation and Readout ... 21
   2.3 Scanning SQUID Microscopy 22
      2.3.1 Control Electronics 22
      2.3.2 Approach Procedure 25
      2.3.3 Vibration Isolation 26
   2.4 Fabrication of Samples 27
   References 28

3 **Results** 29
   3.1 SQUID-on-Tip Characterization 29
      3.1.1 Aluminum SQUIDs 29
      3.1.2 Pb, Nb and Sn SQUIDs 38
   3.2 Imaging 38
      3.2.1 Magnetic Signals from Serpentines:
            Calibration Samples 39

		3.2.2 Vortices in a Niobium Serpentine	42
	References		44
4	**Discussion**		45
	References		46

Appendix A: Explanation of the Negative Differential Resistance 47

Appendix B: Full SOT Circuit Analysis 51

Appendix C: Magnetic Field Profile of a Serpentine 57

Appendix D: Magnetic Field Profile of a Vortex as Seen by the SOT 59

Index 61

Abbreviations

AFM	Atomic force microscopy
BBN	Broad band noise
cP	Centipoise = mPa s
FLL	Flux-locked loop
FWHM	Full width at half maximum
HF	Hydrofluoric acid
MBG	Moving Bragg glass
MTG	Moving transverse glass
NBN	Narrow band noise
NSOM	Near-field scanning optical microscope
Oe	Oersted
PI	Proportional-integral
PID	Proportional-integral-derivative
PLL	Phase locked loop
ROS	Relaxation-oscillation SQUIDs
SC	Superconductor
SEM	Scanning electron microscope
SOT	SQUID on tip
SQUID	Superconducting quantum interference device
SSAA	Series SQUID array amplifier
SSM	Scanning SQUID microscope
TF	Tuning forks
VCO	Voltage controlled oscillator
vdW	van der Waals

Chapter 1
Introduction

1.1 Scientific Background

In the following section I present the scientific background relevant to this work. In Sect. 1.1.1 I review type-II superconductors, concentrating on their dynamical properties in the mixed state, with an emphasis given to vortices and vortex dynamics. In Sect. 1.1.2 a short explanation of how a superconducting quantum interference device (SQUID) works is given and finally in Sect. 1.1.3 I describe the working principle of a tuning fork. Understanding the SQUID and the tuning fork is vital for the study of vortex dynamics.

1.1.1 Vortices in Type-II Superconductors

In 1957, A. A. Abrikosov obtained a spatially periodic square lattice solution for the order parameter in some superconductors [1]. A single flux quantum [2], $\Phi_0 = hc/2e = 2.07 \times 10^{-7}$ Gauss $\times$ cm^2, was associated with each of the periodic zeroes of his solution. He named these zeroes 'vortices' as he noted that the supercurrent curls around them and as such their behavior is analogous to that of vortices in liquid helium. Calling this state the "mixed state", he labeled them as "superconductors of the second group". These are known as type-II superconductors. There are two key parameters, the coherence length, ξ, and the magnetic penetration length, λ, whose ratio, κ, determines whether a superconductor is type-I ($\kappa < 1/\sqrt{2}$) or type-II ($\kappa > 1/\sqrt{2}$). Abrikosov showed that it is energetically favorable for magnetic field to penetrate a type-II superconductor in the form of vortices.

1.1.1.1 Static Vortex Phases

For a type-II superconductor in equilibrium, the mean field prediction is indeed that of an Abrikosov lattice. However, this equilibrium phase diagram is not a simple one, and four different energies compete for dominance, especially in high-T_C superconductors. These four energies are:

1. Thermal energy (thermal disorder)
2. Vortex–vortex interaction
3. Pinning (quenched disorder)
4. Coupling between superconducting layers

The statics of lattices with disorder has been investigated thoroughly in the last two decades, especially in type-II superconductors. Initially, it was agreed that disorder leads to a glass phase called a vortex glass with many metastable states, where barriers between these states become divergent and the positional order at large length scales is lost [3]. This low-temperature phase was expected to be topologically disordered with no positional order. In the scope of this description it was shown that beyond some length scale, the Abrikosov lattice would be destroyed by disorder. However, some experimental observations did not fit this theory, the most important of which was that a first-order transition between the glass phase and the liquid phase was observed at low magnetic fields while at high magnetic fields it was a continuous transition [4–7]. This has led to a different description of the statics of a disordered lattice, predicting a new thermodynamic glassy phase with Bragg diffraction peaks. Hence it was named the Bragg glass [8, 9]. This glass has the following properties [10, 11]: (i) Translational order decays algebraically, i.e. there is quasi-long range order; (ii) it is topologically ordered; (iii) it is still a static glass phase with diverging barriers. The main difference between the vortex glass and the Bragg glass is the existence of the algebraically decaying translational order, manifested as Bragg peaks in, for example, neutron diffraction experiments. A theoretically predicted phase diagram depicting the Bragg glass is presented in Fig. 1.1.

Neutron diffraction experiments confirm the existence of the Bragg peaks [13, 14]. In these experiments the Bragg peaks lose their intensity and become nearly unobservable in the liquid phase as the temperature is increased. Similarly, a loss in their intensity is observed when the magnetic field is increased at a fixed temperature to the vortex glass phase. In addition, magnetization measurements clearly show the first order transition into the liquid phase which becomes a continuous one at higher magnetic fields [7, 15].

1.1.1.2 Dynamic Vortex Phases

If a vortex is in the presence of a current, it will experience a Lorentz force,

$$\mathbf{f} = \mathbf{J}(r) \times \frac{\Phi_0}{c},$$

1.1 Scientific Background

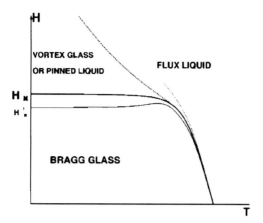

Fig. 1.1 A typical magnetic field–temperature (H-T) phase diagram of a type II superconductor from Ref. [12]

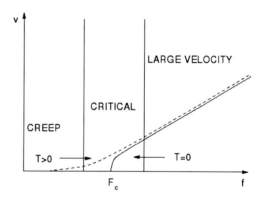

Fig. 1.2 A typical velocity–driving force (v–f) curve at $T = 0$ (*full line*) and $T > 0$ (*dashed line*), taken from Ref. [10]. This can also be termed the "I–V" curve, as the applied current and induced voltage are proportional to the force and average velocity, respectively

where **f** is the force per unit length on the vortex, $\mathbf{J}(r)$ is the current density at r and Φ_0 is parallel to the applied magnetic field [3]. If this force is stronger than the pinning force, the vortex will start moving at some velocity **v**. As a result of this flow of vortices, a finite resistance builds up in the superconductor due to the electric field, $\mathbf{E} = \frac{1}{c}\mathbf{B} \times \mathbf{v}$.

When applying an external force on a periodic vortex lattice, one considers three main regimes, shown in Fig. 1.2:

1. Below the depinning force, f_c, vortices may occasionally move through thermal activation. That is to say, flux lines jump from one pinning point to another to achieve a lower energy configuration. This regime is also known as the flux-creep regime [16]. The theoretical description of this regime is known as the Anderson–Kim flux creep theory, according to which flux creep occurs through jumps of bundles of flux lines. They jump in bundles since the length scale of the repulsion between them is usually larger than the distance between them, which

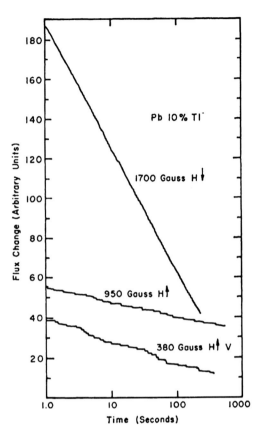

Fig. 1.3 Flux change versus time for different magnetic fields. The steps in the two lower curves are due to digitization in the signal processing unit. Taken from Ref. [17]

facilitates a so-called cooperative behavior. The jump rate, ν, is given by an Arrhenius expression $\nu = \nu_0 e^{-U/k_B T}$, where ν_0 is some characteristic frequency of flux line motion and U is the activation energy, i.e. the height of the barrier for thermally activated motion of a flux bundle. Anderson and Kim derive the flux creep equation for the magnetic field,

$$\frac{\partial B}{\partial t} = \nabla \times \left[\left(\frac{\nabla B}{|\nabla B|} \right) B w \nu_0 e^{-U(B)/k_B T} \right],$$

In the above equation, w is the average distance by which a vortex bundle moves in a thermally activated jump. From this equation they calculate its time dependence, which is logarithmic and shown in Fig. 1.3.

2. Assuming our system indeed behaves like an elastic one, it does not necessarily move under the action of a driving force. The existence of disorder creates a threshold force, f_c, also known as the depinning force. This is a general critical dynamics behavior of such driven interfaces in random media. In the case of vortices it is the result of competition between pinning forces and the vortex

1.1 Scientific Background

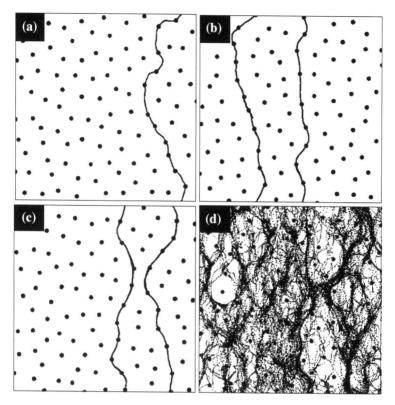

Fig. 1.4 Molecular dynamics simulations of vortex flow in a sample with strong disorder. Taken from Ref. [20]. **a** Traces of vortices throughout the simulation time along with a snapshot of positions. There is a single flow channel, while the rest of the vortices remain pinned. **b** A double channel dynamical state, decoupled. Vortices in each channel flow with different average velocities. **c** A double channel dynamical state, but the two channels interact, resulting in averaging of the velocities in the two channels. **d** Flow in the plastic regime. Note regions with large activity resembling the channels shown in (**a**–**c**) and regions with small finite activity elsewhere. A vortex in this flow will participate alternately in channel-like flow or be pinned in a pinning site which is outside of the channel

lattice's elastic properties. Near the depinning transition, $f \approx f_c$, and in strong disorder depinning is observed to proceed via the flowing of vortices through plastic channels [18, 19]. Molecular dynamic simulations [20] take into account both the vortex interaction energy $U_{vv}(r_{ij})$ and the pinning potential $U_{vp}(r) = -A_p e^{-(r/a_p)^2}$, where a_p is the interaction range of impurity and A_p is the pinning strength. The corresponding equation of motion (normalized) is

$$\frac{d\mathbf{r}_i}{dt} = -\sum_{j \neq i} \nabla_i U_{vv}(r_{ij}) - \sum_k \nabla_i U_{vp}(r_{ik}) + \mathbf{f} + \eta_i(t).$$

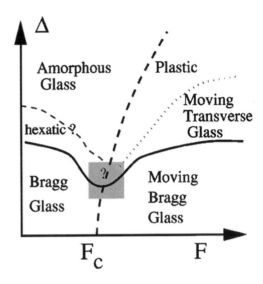

Fig. 1.5 A schematic phase diagram from Ref.[10] for $d = 3$ at zero temperature. Here Δ is the level of disorder and F is the driving force. The behavior in the *square gray* region is unclear

Here **f** is the force on the vortices and η is thermal noise. An illustrative result of these simulations is shown in Fig. 1.4.

In addition, using Lorentz microscopy, Matsuda et al. managed to record on video this plastic flow [21]. In this video, they show how vortices move in channels, or "rivers", and at some point this channel-like motion turns into motion of the entire vortex lattice (elastic flow).

3. Well above the depinning force, moving vortices seem to be more translationally ordered than at lower velocities. As was shown in Ref. [10], some modes of the disorder are not affected by the motion even at large velocities and as a result the vortices exist in a phase called a moving Bragg glass (MBG). In this phase, vortex flow is in the form of channels, which are aligned almost parallel to each other and these channels are coupled along the axis perpendicular to their motion. They are sometimes called "coupled channels" or "elastic channels". Once these channels are established, it is energetically unfavorable for them to reorient. Therefore, there exists some critical transverse force below which the transverse motion of the vortices is pinned. The level of disorder affects the MBG so that with strong enough disorder, the system goes from the MBG to a moving transverse glass [22, 23] (MTG). This phase is also called a smectic flow [24] since the channels themselves become the elementary particles and is associated with the decoupling of the channels, while periodicity in the transverse direction is maintained. A schematic diagram taken from Ref. [10] is given in Fig. 1.5.

An important feature of the MBG is the existence of a sharp frequency associated with the periodicity of the lattice, also known as the washboard frequency. This frequency exists for a lattice moving at an average velocity v with a lattice constant a, $\nu = v/a$.

1.1 Scientific Background

1.1.2 Superconducting Quantum Interference Device (SQUID)

A short derivation of the Josephson equations [25] is given below. It was first suggested by Feynman [26] and is repeated here due to its clear and simple explanation of the effect, focusing on magnetic field dependence properties. For notations, a superconducting electrode can be written as a macroscopic wave function of the form $\psi = \sqrt{\rho} e^{i\varphi}$, where $\rho = |\psi|^2$. As usual, the electric current density in the presence of a vector potential can be written as

$$\mathbf{J} = \frac{e^*}{m^*} \left[\frac{i\hbar}{2} \left(\psi \nabla \psi^* - \psi^* \nabla \psi \right) - \frac{e^*}{c} \mathbf{A} |\psi|^2 \right],$$

where $e^* = 2e$ and $m^* = 2m$. With the expression above for ψ, $\mathbf{J}$ becomes

$$\mathbf{J} = \rho \frac{e}{m} \left(\hbar \nabla \varphi - \frac{2e}{c} \mathbf{A} \right). \tag{1.1}$$

Next we consider two coupled superconducting electrodes, in which each electrode's wave function has its tail entering the other (decaying). We denote the left electrode's wave function as ψ_L and the right as ψ_R, so that with a coupling potential K between them, the Schrödinger equations for the junction are

$$i\hbar \dot{\psi}_R = \mathcal{H}_R \psi_R + K \psi_L = E_R \psi_R + K \psi_L$$

$$i\hbar \dot{\psi}_L = \mathcal{H}_L \psi_L + K \psi_R = E_L \psi_L + K \psi_R.$$

For each electrode, $E_{L,R} = 2\mu_{L,R}$ (twice the electrochemical potential), as it is the minimum energy required to add a Cooper pair to the system. If there is a potential difference, V, across the junction, then we can choose the zero of the energy such that

$$i\hbar \dot{\psi}_R = -eV \psi_R + K \psi_L$$

$$i\hbar \dot{\psi}_L = eV \psi_L + K \psi_R.$$

Now we substitute the expressions for the wave functions and separate real and imaginary terms to get four equations:

$$\frac{\partial \rho_L}{\partial t} = \frac{2}{\hbar} K \sqrt{\rho_L \rho_R} \sin \varphi \tag{1.2}$$

$$\frac{\partial \rho_R}{\partial t} = -\frac{2}{\hbar} K \sqrt{\rho_L \rho_R} \sin \varphi \tag{1.3}$$

$$\frac{\partial \varphi_L}{\partial t} = \frac{K}{\hbar} K \sqrt{\rho_L / \rho_R} \cos \varphi + eV/\hbar \tag{1.4}$$

$$\frac{\partial \varphi_R}{\partial t} = \frac{K}{\hbar} K\sqrt{\rho_L/\rho_R} \cos\varphi - eV/\hbar, \qquad (1.5)$$

with $\varphi = \varphi_L - \varphi_R$.
From the first two equations, the pair current density is

$$J = \partial \rho_L/\partial t = \frac{2K}{\hbar}\sqrt{\rho_L \rho_R} \sin\varphi \equiv J_1 \sin\varphi.$$

This is known as the DC Josephson equation. From the last two equations,

$$\frac{\partial \varphi}{\partial t} = \frac{2eV}{\hbar},$$

which is known as the AC Josephson equation. To understand the junction's behavior under magnetic field, we recall Eq. 1.1 and rewrite it as

$$\nabla\varphi = \frac{2e}{\hbar c}\left(\frac{mc}{2e^2 \rho}\mathbf{J} + \mathbf{A}\right).$$

We follow the derivation of Ref. [27] and write the phase's dependence on magnetic field

$$\varphi = \frac{2e}{\hbar c} d H_y x + \varphi_0,$$

so that

$$J = J_1 \sin\left(\frac{2e}{\hbar c} d H_y x + \varphi_0\right).$$

In this derivation, d is the effective magnetic penetration length of the two junctions combined and the magnetic field is in the y-direction and the junction lies along the x-direction, so that the integration is performed along it in order to find the contribution of the entire cross section of the junction. This result shows that the critical current density is spatially modulated by the magnetic field in a periodic way. One can then integrate along the junction's cross-section and find the known dependence on magnetic field.

A more interesting case is that of a superconducting ring interrupted by two Josephson junctions, also known as a Superconducting QUantum Interference Device (SQUID). Relating to Fig. 1.6, we draw a contour along the ring, and to simplify the derivation, we assume that the contour is deep enough inside the superconductors such that $\mathbf{J} = 0$ or $\nabla\varphi = \frac{2e}{\hbar c}\mathbf{A}$ along the contour. Next, we integrate this relation along the contour. Note that the two junctions need to be taken into consideration:

$$[\varphi_R(1) - \varphi_L(1)] - [\varphi_R(2) - \varphi_L(2)] = \frac{2e}{\hbar c} \oint \mathbf{A} \times \mathbf{dl},$$

or, written more elegantly using $\gamma_1 \equiv \varphi_R(1) - \varphi_L(1)$ and $\gamma_2 \equiv \varphi_R(2) - \varphi_L(2)$:

1.1 Scientific Background

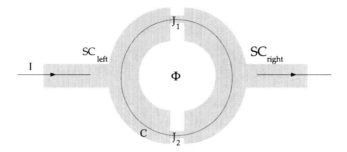

Fig. 1.6 A schematic of a SQUID with microbridges (constrictions) serving as the Josephson junctions. The two bulky superconducting electrodes are marked as SC$_{left}$ and SC$_{right}$ and the junctions as J$_1$ and J$_2$. The integration contour is labeled with the letter C

$$\gamma_1 - \gamma_2 = \frac{2e}{\hbar c} \oint \mathbf{A} \cdot d\mathbf{l} = 2\pi \frac{\Phi}{\Phi_0}. \quad (1.6)$$

Here Φ is the total flux through the loop. With this result we can now write the total current as a sum of the currents through each junction:

$$I = I_0 \sin \gamma_1 + I_0 \sin \gamma_2 = 2I_0 \cos\left(\pi \frac{\Phi}{\Phi_0}\right) \sin\left(\gamma_2 + \pi \frac{\Phi}{\Phi_0}\right) \quad (1.7)$$

with a maximum current of

$$I = 2I_0 \left|\cos\left(\pi \frac{\Phi}{\Phi_0}\right)\right|.$$

This result can be used to calculate the magnetic flux through a loop by measuring the critical current. My master's thesis was dedicated to the initial development of a submicron SQUID on a tip [28]. This prototype SQUID-on-tip (SOT) had a loop diameter of several hundred nanometers and a noise sensitivity of 50 mG/$\sqrt{\text{Hz}}$. A typical I–H curve of this kind of SQUID is given in Fig. 1.7. A detailed analysis of the equations of motion governing our SQUID is given in Appendix B.

1.1.3 Tuning Fork Microscopy

When a scanning probe, coupled to a tuning fork, approaches the surface of a sample, elastic and dissipative forces shift its (the tuning fork plus the tip) resonant frequency. Mainly due to availability, accuracy and size, most groups use a tuning fork similar to the component of many quartz-based watches, i.e. one whose resonant frequency is 32 kHz [29, 30]. At distances smaller than 10 nm, the frequency shift due to surface forces is rather large (about 100 mHz per nm) and is possible to detect with a

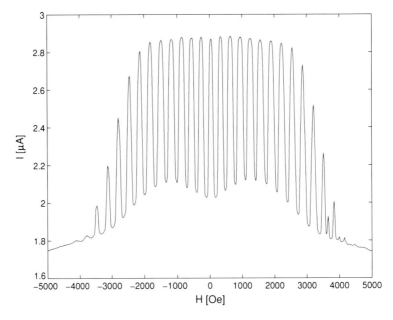

Fig. 1.7 A typical I–H curve of a SQUID on a tip with a diameter of 300 nm

phase-locked loop (PLL). With this detection scheme, it is possible to use the tuning fork as a topography sensor for scanning probe microscopy.

1.1.3.1 The Harmonic Oscillator Approximation

We employ quartz tuning forks (TFs), widely used in wrist watches as their time base. In Fig. 1.8(left) we show an image of a tip lying against a tuning fork just before gluing the two together. The most popular TFs are those with a resonant frequency of 2^{15} or 32,768 Hz and each prong's dimensions are $l \times w \times t = 4 \times 0.34 \times 0.6$ mm^3. We performed a finite element analysis (FEA) for a single crystal quartz tuning-fork of the same dimensions with a small tip glued to its end. The input to this program is a three-dimensional drawing of the tuning-fork having the same dimensions as mentioned above and the Young's modulus for quartz. The output are the eigenmodes of the tuning fork, where we pick only the eigenfrequency of 32,553 Hz (lower modes are canceled by the electrical contacts). This mode is displayed in Fig. 1.8 (right). The displacement is exaggerated for clarity. From its dimensions (length, width, thickness), density, ρ, and Young's modulus, E, one can calculate [31] the spring constant $k = (E/4)w(t/l)^3 = 25{,}320$ N/m. In order to get the proper eigenmodes and eigenfrequencies of the tuning fork, one should write the equations of motion for the two prongs and the base and solve. Phenomenologically, however, it can be described [31] with just one harmonic oscillator equation of motion for the position,

1.1 Scientific Background

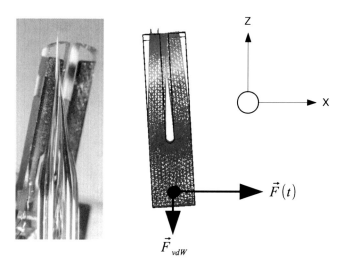

Fig. 1.8 *Left* An optical microscope image of a tip lying against a tuning-fork. *Right* a finite element analysis of the tuning-fork with a tip glued to it. The simulation yields an eigenfrequency of 32,553 Hz. The displacement shown is exaggerated in order the clarify the mode of oscillation

$x(t)$, of one of the prongs and then bear in mind that there are in fact two of them,

$$m\ddot{x} + m\gamma\dot{x} + kx = F(t)$$

Here m is the mass of the prong, γ represents the losses (dissipation) and $F(t)$ is the driving force, in this case represented as a mechanical, dither-type, excitation. The amplitude has, of course, the form of a Lorentzian with a resonant frequency $f_0 = (1/2\pi)\sqrt{k/m}$ and FWHM $= \Delta f \equiv \sqrt{3}\gamma/2\pi$. We also define the quality factor as $Q \equiv f_0/\Delta f$.

1.1.3.2 Tip-Sample Interaction

As mentioned by Giessibl [32], the interaction of a macroscopic tip with a sample is a highly complex many-body problem, so that writing a simple expression for the force between them, F_{ts}, is not simple. Typically, two dominant contributions to F_{ts} are the long-range and short-range interactions, which can be approximated by van der Waals and Morse-type potentials, respectively, so that one can write,

$$F_{ts}(z) = \frac{C}{z+\sigma} + 2\kappa E_{\text{bond}}\left[-e^{-\kappa(z-\sigma)} + e^{-2\kappa(z-\sigma)}\right]$$

C depends on the tip angle. E_{bond}, σ and κ are the bonding energy, equilibrium distance and decay length of the Morse potential, respectively. A detailed calculation can be found in Ref. [32] and references therein. What *is* important is that this force is a function of the potential energy between the tip and the sample, resulting in an effective spring constant between the two, $k_{ts} = -\partial F_{ts}/\partial z$. When the tip is far away from the sample, the resonant frequency is $2\pi f_0 = \sqrt{k_0/m}$. Assuming k_{ts} is constant during the oscillation cycle and that the change is small, i.e. $k_{ts} \ll k_0$, the resonant frequency close to the sample changes to $f = f_0 + \Delta f = \sqrt{k/m}$ with $k = k_0 + k_{ts}$. Using a Taylor expansion we get that the frequency shift is

$$\Delta f = f_0 \frac{k_{ts}}{2k}$$

Since in a tuning fork we actually treat only one prong in our calculation whereas one should take both into consideration, we need to insert a factor of two to the expression,

$$\Delta f_{\text{tuning fork}} = f_0 \frac{k_{ts}}{4k} \tag{1.8}$$

More generally, if k_{ts} is not constant during an oscillation cycle, the calculation is more complicated and can be calculated using perturbation theory [33] or with a straight-forward calculation using the equation of motion [34], so that

$$\Delta f(\vec{x}) = \frac{f_0}{4k} \langle k_{ts} \rangle (\vec{x}) \tag{1.9}$$

with an average spring constant

$$\langle k_{ts} \rangle (\vec{x}) = \frac{2}{\pi} \int_{-1}^{1} k_{ts}(\vec{x} + \vec{e}_\theta \zeta A) \sqrt{1 - \zeta^2} \, d\zeta. \tag{1.10}$$

Here the unit vector $e_\theta = (0, \cos\theta, \sin\theta)$ points in the direction of the oscillation and A is the amplitude of oscillation. In our case, the tuning-fork is excited so that it vibrates almost parallel to the sample [35]. Figure 1.9(left) shows the geometry of such a setup, also known as shear-force microscopy. In Fig. 1.9(right) a representative Δf versus z plot from our own system is given. In agreement with others [36–38], the tuning fork "feels" the sample only a few tens of nanometers away and is mostly insensitive to it when it is farther than that. We also note that the typical frequency shift is approximately 500 mHz.

1.2 Open Questions

Current experimental evidence in the high velocity regimes of vortex matter, i.e. the MBG and MTG, is not very extensive. There were attempts to search for MBG using muon spin rotation and small angle neutron diffraction [39]. Channel flow (both

1.2 Open Questions

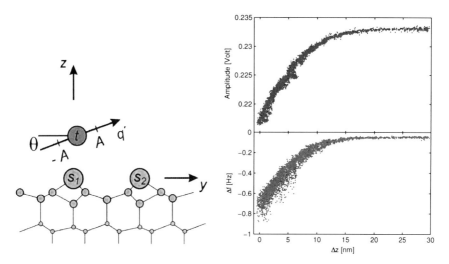

Fig. 1.9 *Left* Geometry of the tip (t) atom and the sample atoms (s_1, s_2). The tip oscillates along an axis that is tilted by an angle θ with respect to the y axis. Taken from Ref. [35]. *Right* Measurement of the decrease in amplitude and shift in resonant frequency of a quartz tuning fork glued to a tip as a function of the distance of the tip's edge from the surface of a sample

smectic and elastic) has been studied in decoration experiments [40, 41]. These results, however, are not conclusive and further experiments are needed to confirm the observations.

The washboard frequency has been observed directly in the creep and plastic regimes using an STM [42]. This STM experiment, however local it is, is limited in both its scanning area and its "shutter speed". Togawa et al. [43] were able to observe it macroscopically in the creep regime in a transport experiment. They attributed the observed broad-band noise (BBN) at low magnetic fields to the plastic flow while a narrow-band noise (NBN) associated with the washboard frequency appeared at high fields and the BBN decreased in amplitude. As they increased the magnetic field even further, the NBN's amplitude decreased and it lost its sharpness, i.e. its narrow-band characteristic. This has not yet been fully explained.

Our scanning nanoSQUID-on-tip microscope features a magnetic field sensitivity on par with existing techniques and yet has a smaller sensor size. In terms of bandwidth it is only limited by the electronics used to read our cryogenic current-voltage converter, currently rated at 1 MHz. Most importantly, it is the only true SQUID, which can be brought safely to within a few nanometers of the sample, thus compensating for signal attenuation due to sensor-sample separation. Using the scanning nanoSQUID microscope presented below in Sect. 2.3, we can now explore the MBG/MTG phase space of the dynamic phase diagram by *locally* measuring the characteristics of the washboard frequency in the different phases. This should manifest itself in the spectral density of the noise in the magnetic signal of our sensor. We can then compare to existing data to see whether there is agreement with previous experiments (low frequencies) and to better understand the mechanism behind the different dynamical phases by analyzing our data for the entire measurable spectrum.

1.3 Goal

- *Characterization of the SQUID-on-tip.* First invented in 2006 [28], the SQUID-on-tip (SOT) has not been studied extensively until now. Specifically, we wish to understand its peculiar current-voltage characteristics and quantify its properties, e.g. magnetic penetration length, inherent noise figure.
- *Construction of a scanning SQUID microscope.* In order to locally study vortices in superconductors, we are currently constructing a scanning probe microscope (SPM) with the "SQUID on a tip" as the magnetic field sensor attached to a tuning fork, which acts as the feedback mechanism. This SQUID is considerably smaller than typical SQUIDs used in scanning probe microscopes and it also has the advantage of being on the edge of a pointy probe, which makes it a true local sensor. In addition, its relatively low resistance (hundreds of Ohms) makes high-frequency (more than 100 kHz) measurements plausible. Once the microscope is assembled, we plan to first test it on a known crystal and observe the statics of a vortex lattice, e.g. 2H-NbSe$_2$ or BSCCO.
- *Imaging vortices in type-II superconductors in the static and dynamic regimes.* After the initial assembly of the microscope is complete, we wish to study the behavior of vortices in type-II superconductors in both the static regime as a gauge of the microscope quality and in the dynamic regime over a wide range of velocities in order to study vortex dynamics in the plastic flow regime.

References

1. Abrikosov, A. A. On the magnetic properties of superconductors of the second group. *Sov. Phys. JETP* **5**, 1174 (1957).
2. Deaver, B. S. and Fairbank, W. M. Experimental evidence for quantized flux in superconducting cylinders. *Phys. Rev. Lett.* **7**, 43 (1961).
3. Blatter, G., Feigelman, M. V., Geshkenbein, V. B., Larkin, A. I., and Vinokur, V. M. Vortices in high-temperature superconductors. *Rev. Mod. Phys.* **66**, 1125 (1994).
4. Kwok, W. K., Fleshler, S., Welp, U., Vinokur, V. M., Downey, J., Crabtree, G. W., and Miller, M. M. Vortex lattice melting in untwinned and twinned single crystals of YBa$_2$Cu$_3$O$_{7-\delta}$. *Phys. Rev. Lett.* **69**, 3370 (1992).
5. Charalambous, M., Chaussy, J., and Lejay, P. Evidence from resistivity measurements along the c axis for a transition within the vortex state for **H** ∥ ab in single-crystal YBa$_2$Cu$_3$O$_7$. *Phys. Rev. B* **45**, 5091 (1992).
6. Safar, H., Gammel, P. L., Huse, D. A., Bishop, D. J., Lee, W. C., Giapintzakis, J., and Ginsberg, D. M. Experimental evidence for a multicritical point in the magnetic phase diagram for the mixed state of clean untwinned YBa$_2$Cu$_3$O$_7$. *Phys. Rev. Lett.* **70**, 3800 (1993).

References

7. Zeldov, E., Majer, D., Konczykowski, M., Geshkenbein, V. B., Vinokur, V. M., and Shtrikman, H. Thermodynamic observation of first-order vortex-lattice melting transition in $Bi_2Sr_2CaCu_2O_8$. *Nature* **375**, 373 (1995).
8. Giamarchi, T. and Le Doussal, P. Elastic theory of pinned flux lattices. *Phys. Rev. Lett.* **72**, 1530 (1994).
9. Giamarchi, T. and Le Doussal, P. Elastic theory of flux lattices in the presence of weak disorder. *Phys. Rev. B* **52**, 1242 (1995).
10. Le Doussal, P. and Giamarchi, T. Moving glass theory of driven lattices with disorder. *Phys. Rev. B* **57**, 11356 (1998).
11. Nattermann, T. and Scheidl, S. Vortex-glass phases in type-II superconductors. *Adv. Phys.* **49**, 607 (2000).
12. Giamarchi, T. and Le Doussal, P. Phase diagrams of flux lattices with disorder. *Phys. Rev. B* **55**, 6577 (1997).
13. Yaron, U., Gammel, P. L., Huse, D. A., Kleiman, R. N., Oglesby, C. S., Bucher, E., Batlogg, B., Bishop, D. J., Mortensen, K., Clausen, K., Bolle, C. A., and De La Cruz, F. Neutron diffraction studies of flowing and pinned magnetic flux lattices in $2H\text{-}NbSe_2$. *Phys. Rev. Lett.* **73**, 2748 (1994).
14. Cubitt, R., Forgan, E. M., Yang, G., Lee, S. L., Paul, D. M., Mook, H. A., Yethiraj, M., Kes, P. H., Li, T. W., Menovsky, A. A., Tarnawski, Z., and Mortensen, K. Direct observation of magnetic flux lattice melting and decomposition in the high-T_c superconductor $Bi_{2.15}Sr_{1.95}CaCu_2O_{8+x}$. *Nature* **365**, 407 (1993).
15. Beidenkopf, H., Avraham, N., Myasoedov, Y., Shtrikman, H., Zeldov, E., Rosenstein, B., Brandt, E. H., and Tamegai, T. Equilibrium first-order melting and second-order glass transitions of the vortex matter in $Bi_2Sr_2CaCu_2O8$. *Phys. Rev. Lett.* **95**, 257004 (2005).
16. Anderson, P. W. Theory of flux creep in hard superconductors. *Phys. Rev. Lett.* **9**, 309 (1962).
17. Beasley, M. R., Labusch, R., and Webb, W. W. Flux creep in type-II superconductors. *Phys. Rev.* **181**, 682 (1969).
18. Brandt, E. H. Computer simulation of flux pinning in type-II superconductors. *Phys. Rev. Lett.* **50**, 1599 (1983).
19. Jensen, H. J., Brass, A., and Berlinsky, A. J. Lattice deformations and plastic flow through bottlenecks in a two-dimensional model for flux pinning in type-II superconductors. *Phys. Rev. Lett.* **60**, 1676 (1988).
20. Grønbech-Jensen, N., Bishop, A. R., and Domínguez, D. Metastable filamentary vortex flow in thin film superconductors. *Phys. Rev. Lett.* **76**, 2985 (1996).
21. Matsuda, T., Harada, K., Kasai, H., Kamimura, O., and Tonomura, A. Observation of dynamic interaction of vortices with pinning centers by Lorentz microscopy. *Science* **271**, 1393 (1996). http://rdg.ext.hitachi.co.jp/rd/moviee/vortices2-n.mpeg.
22. Berthier, C., Levy, L., and Martinez, G., editors. *High Magnetic Fields: Applications in Condensed Matter Physics and Spectroscopy*, volume 595 of *Lecture Notes in Physics*. Springer, (2002). p. 314.
23. Moon, K., Scalettar, R. T., and Zimányi, G. T. Dynamical phases of driven vortex systems. *Phys. Rev. Lett.* **77**, 2778 (1996).
24. Balents, L., Marchetti, M. C., and Radzihovsky, L. Nonequilibrium steady states of driven periodic media. *Phys. Rev. B* **57**, 7705 (1998).
25. Josephson, B. D. Possible new effects in superconductive tunnelling. *Physics Letters* **1**, 251 (1962).
26. Feynman, R. P., Leighton, R. B., and Sands, M. *The Schrödinger equation in a classical context: A seminar on superconductivity*, volume III of *The Feynman Lectures on Physics*. Addison-Wesley, (1965).
27. Fossheim, K. and Sudbø, A. *Superconductivity Physics and Applications*, chapter 5, 123–140. John Wiley & Sons, Ltd (2005).
28. Finkler, A. A SQUID on a tip: A tool to explore vortex matter in high-T_C superconductors. M.Sc. thesis, Weizmann Institute of Science, Rehovot, Israel, November (2006).

29. Grober, R. D., Acimovic, J., Schuck, J., Hessman, D., Kindlemann, P. J., Hespanha, J., Morse, A. S., Karrai, K., Tiemann, I., and Manus, S. Fundamental limits to force detection using quartz tuning forks. *Rev. Sci. Instrum.* **71**, 2776 (2000).
30. Rychen, J., Ihn, T., Studerus, P., Herrmann, A., and Ensslin, K. A low-temperature dynamic mode scanning force microscope operating in high magnetic fields. *Rev. Sci. Instrum.* **70**, 2765 (1999).
31. Karrai, K. and Grober, R. D. Piezo-electric tuning fork tip-sample distance control for near field optical microscopes. *Ultramicroscopy* **61**, 197 (1995).
32. Giessibl, F. J. Advances in atomic force microscopy. *Rev. Mod. Phys.* **75**, 949 (2003).
33. Giessibl, F. J. Forces and frequency shifts in atomic-resolution dynamic-force microscopy. *Phys. Rev. B* **56**, 16010 (1997).
34. Dürig, U. Relations between interaction force and frequency shift in large-amplitude dynamic force microscopy. *Surf. Interface Anal.* **75**, 433 (1999).
35. Giessibl, F. J., Herz, M., and Mannhart, J. Friction traced to the single atom. *Proc. Nat. Acad. Sci.* **99**, 12006 (2002).
36. Karrai, K. and Grober, R. D. Piezoelectric tip-sample distance control for near field optical microscopes. *Appl. Phys. Lett.* **66**, 1842 (1995).
37. Seo, Y., Jhe, W., and Hwang, C. S. Electrostatic force microscopy using a quartz tuning fork. *Appl. Phys. Lett.* **80**, 4324 (2002).
38. Atia, W. A. and Davis, C. C. A phase-locked shear-force microscope for distance regulation in near-field optical microscopy. *Appl. Phys. Lett.* **70**, 405 (1997).
39. Charalambous, D., Kealey, P. G., Forgan, E. M., Riseman, T. M., Long, M. W., Goupil, C., Khasanov, R., Fort, D., King, P. J. C., Lee, S. L., and Ogrin, F. Vortex motion in type-II superconductors probed by muon spin rotation and small-angle neutron scattering. *Phys. Rev. B* **66**, 054506 (2002).
40. Marchevsky, M., Aarts, J., Kes, P. H., and Indenbom, M. V. Observation of the correlated vortex flow in $NbSe_2$ with magnetic decoration. *Phys. Rev. Lett.* **78**, 531 (1997).
41. Pardo, F., de la Cruz, F., Gammel, P. L., Bucher, E., and Bishop, D. J. Observation of smectic and moving-bragg-glass phases in flowing vortex lattices. *Nature* **396**, 348 (1998).
42. Troyanovski, A. M., Aarts, J., and Kes, P. H. Collective and plastic vortex motion in superconductors at high flux densities. *Nature* **399**, 665 (1999).
43. Togawa, Y., Abiru, R., Iwaya, K., Kitano, H., and Maeda, A. Direct observation of the washboard noise of a driven vortex lattice in a high-temperature superconductor, $Bi_2Sr_2CaCu_2O_y$. *Phys. Rev. Lett.* **85**, 3716 (2000).

Chapter 2
Methods

The experimental setup used in this work consists of three major subsystems. The first and foremost in importance is the SQUID-on-tip. In Sect. 2.1 we describe the procedure used to fabricate it. The second subsystem is the tuning-fork assembly, which is described in detail in Sect. 2.2. These two subsystems comprise the sensor assembly of the scanning SQUID microscope. Their integration into one assembly is described in Sect. 2.3. The third and last subsystem is the fabrication of samples. I will give a brief yet thorough description of this procedure in Sect. 2.4.

2.1 SQUID-on-Tip Fabrication

We pull a quartz tube with an outer diameter of 1 mm and inner diameter of 0.5 mm to form a pair of sharp pipettes with a tip diameter that can be controllably varied between 100 and 400 nm using a commercial pipette puller.[1] Then, we either apply a thin layer of indium or deposit a thin (200 nm) film of gold on two sides of the cylindrical base of the pipette. Afterwards, the pipette is mounted on a rotator (see Fig. 2.1) and put into a vacuum chamber for three steps of thermal evaporation of aluminum. The rotator has an electrical feedthrough for two operations. First, a 2 mW red laser diode is connected, which allows for the alignment of the tip main axis with respect to the source. This defines our zero angle. The second electrical connection is for an in-situ measurement of the tip's resistance during deposition. A finite resistance typically appears after depositing approximately 70 Å in the third stage (see below). In the first step (see Fig. 2.2), 25 nm of aluminum are deposited on the tip tilted at an angle of $-100°$ with respect to the line to the source. Then the tip is rotated to an angle of $100°$, for a second deposition of 25 nm. As a result, two leads on opposite sides of the quartz tube, reaching all the way to the apex, are formed. In the last step 20–22 nm of aluminum are deposited at an angle of $0°$, coating the apex ring of the tip. The resulting structure has two leads connected to a ring. "Strong" superconducting regions are formed in the areas where the leads make contact with

---
[1] Sutter Instruments P-2000.

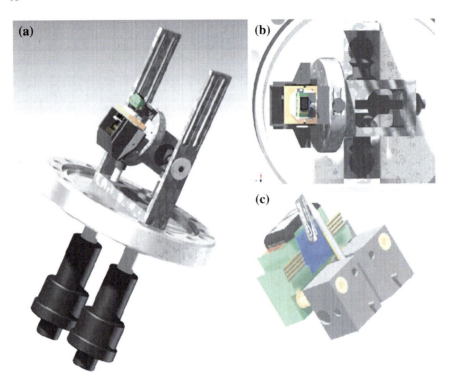

Fig. 2.1 a A CAD image of the rotator used to rotate tips in the vacuum chamber. The original design of this rotator was based on a similar device from Yoo et al. [1]. **b** A close-up view of the *gear-box* holding the rotating stage with the tip holder. **c** The tip holder, co-linear with the laser diode on the rotator. The tuning-fork assembly is not present during deposition for practical reasons

the ring, while the two parts of the ring in the region of the gap between the leads constitute two weak links, thus forming the SQUID. Its typical room-temperature resistance is 1.5 kΩ. Since the device is highly sensitive to electrostatic discharge, it is shorted when being transferred from the evaporator to the microscope.

2.2 Tuning Fork Assembly

The idea to use a quartz tuning-fork as a topography sensor was suggested to us by Dr. Michael Rappaport. He had previously built a tuning-fork based NSOM [2], and the basic principles of operation were taken from his fiber-optic design and adapted to our much larger SQUID-on-tip.

2.2 Tuning Fork Assembly

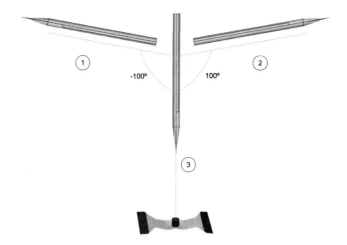

Fig. 2.2 A schematic of the three-stage thermal deposition of a SOT

2.2.1 Preparation

We use commercial[2] quartz tuning forks, laser-trimmed to have a resonance at 2^{15} Hz, which are normally used as time bases in digital watches. Preparing one for microscopy entails the removal of the vacuum can and the two electrodes soldered to it (see Fig. 2.3). We then glue it to a $5 \times 5 \times 0.5$ mm^3 quartz piece which was pre-coated with two gold electrodes (70 Å Cr/2000 Å Au). The two tuning-fork electrodes are bonded to those quartz electrodes, which are in turn connected to external wires via two additional bonds. This allows for the electrical readout of the voltage, which develops on the tuning-fork due to the piezoelectric effect. In principle this would have been enough, since it is possible to electrically excite the tuning-fork with a voltage source and read the current through it using a current-to-voltage converter [3]. However, it is more advantageous to decouple the excitation from the readout by driving voltage to a dither piezo, which mechanically excites the tuning-fork (Fig. 2.3d). We used two types of dither piezos for this purpose, one of which[3] had the same dimensions as the quartz plate and was therefore used both as the plate to which we glued the tuning-fork and as its mechanical excitation medium. The second[4] was slightly bulkier and was therefore placed on the side of the tuning-fork holder.

---

[2] Polaros Electronics, TB38-20-12.5-32.768 KHz.
[3] EBL#4.
[4] PI PICMA PL033.30.

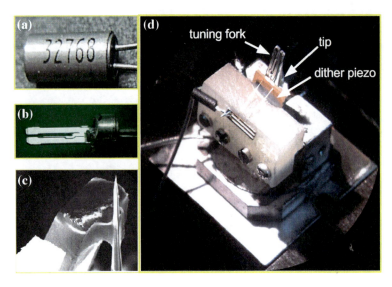

Fig. 2.3 a A commercial tuning fork encased in a vacuum can. **b** The same tuning fork after the vacuum can had been removed. **c** A SEM micrograph of a tip glued to a prong of a tuning fork. **d** The tip/tuning fork assembly. The tuning fork itself is glued onto the dither piezo with E32 and then glued to the tip with Araldite 2020

2.2.2 Gluing a Tip and Its Effect on Resonance

In our microscope setup, we glue the SOT to one of the prongs of the TF (see Fig. 2.3c). This can change the resonant frequency [4] dramatically. Therefore, we try to use as small amount of glue as possible. This is made possible by using a two-part epoxy with a rather low (150 cP) viscosity, which enables us to place a very small drop of glue on the part joining the two. The longer the SOT protrudes above the edge of the prong, the softer its effective spring constant [5] becomes (for a cylindrical tip)

$$k_{\text{eff}} = \frac{3\pi E r^4}{4l^3}$$

Here r, l and E are the radius, length and Young modulus of the tip, respectively. Thus, if it protrudes too much, the vdW forces of interaction with the sample will dampen its motion before the tuning fork's prong actually "feels" anything. Consequently, we always try to position the tip with respect to the tuning fork so that the SOT itself protrudes only slightly above the edge of the prong (typically a few tens of microns). We then excite the tuning fork and measure its resonance, specifically noting the location of the peak, its amplitude and its width. This measurement is performed at room temperature and atmospheric pressure and repeated at low temperature (300 mK) in vacuum for comparison. More often than not, the low-

2.2 Tuning Fork Assembly

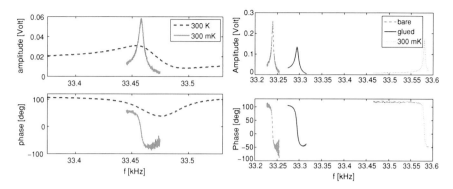

Fig. 2.4 The resonance of our quartz tuning fork. *Left* comparison between the Q-factor at room temperature and at low temperature. For ease of comparison, the latter was shifted by 80 Hz so that its peak overlaps with that of the former; *Right* the "bare" and "glued" curves are at room temperature, with "bare" meaning that the tuning-fork is not glued to the tip. The 300 mK curve is that of the "glued" tip cooled down to 300 mK

temperature, vacuum resonance is much sharper than the room-temperature, ambient pressure one (see Fig. 2.4).

2.2.3 Mechanical Versus Electrical Excitation and Readout

In the equivalent electrical model for a resonating tuning-fork [6, 7], there are four parameters which should be considered: the stray capacitance, $C_0 \sim pF$ and the RLC-equivalent parameters, which in our case are $R = 30\,k\Omega, C = 1\,fF, L = 23.6\,kH$. The stray capacitance is in fact that of the contacts. This means that the tuning-fork is in effect a reactive impedance, with an effective impedance of $\approx 1\,M\Omega$. Any additional stray capacitance, mostly wires, will attenuate the signal considerably.

There are two methods of exciting and measuring the resonance of the tuning-fork:

1. Apply a voltage between the two prongs of the tuning-fork and measure the current through it. This is a compact method since it requires as little as two wires. The disadvantage with such a method is that there is coupling between the electrical excitation (piezoelectric effect) and the current readout, giving rise to a large background signal. In order to overcome this, one can insert a variable capacitor and a center-tapped transformer [3], which can be varied so as to exactly cancel the stray capacitance.
2. Mechanically excite the tuning-fork using an external dither piezo and reading the voltage between the two prongs. This way, the excitation is electrically decoupled from the signal readout. There still remains, however, the problem of signal attenuation due to cables. Once again one may employ the variable capacitor method. Another, more complicated but by far superior solution, is to use a

trans-impedance converter right next to the tuning-fork before the signal is attenuated by the experimental setup's cables, so that the effective impedance is converted from a few MΩ to a few kΩ.

We used both techniques in our setup. The trans-impedance converter we used was made by atto**cube**, and comprised of a commercial GaAs MESFET, which can work at cryogenic temperatures (charge carriers in GaAs do not freeze as those in silicon do).

2.3 Scanning SQUID Microscopy

The microscope's design is based on commercial coarse positioners and piezoelectric scanners from atto**cube**. It consists of two parts: the bottom flange, which holds the coarse-z positioner, the xyz scanner and the sample; and the top flange, which holds the coarse-x and y positioners and the tip holder assembly (see Fig. 2.5). Our microscope's feedback mechanism relies on the behavior of a quartz tuning fork's (TF) resonance as it is moves closer and closer to the surface of a sample. Both the quartz TF's modes and the above-mentioned behavior have been studied extensively and are widely used today in near-field scanning optical microscopes (NSOM) and atomic force microscopes (AFM) [8, 9]. Our tuning-forks are laser-trimmed so that they peak at $2^{15} = 32,768$ Hz.

2.3.1 Control Electronics

We divide the electronics section into three parts: (a) the SQUID on tip; (b) the tuning fork; and (c) coarse and fine motion.

2.3.1.1 SQUID on Tip Electronics

A SQUID is by definition a low impedance device. Typically, room-temperature electronics are optimized for high impedances. Consequently, this impedance mismatch results in a decrease in the signal-to-noise ratio. There are many schemes which tackle this problem successfully, e.g. non-dispersive read-out [10] and transformer impedance matching [11]. We chose to use a SQUID series array amplifier [12], which is in essence an array of 100 SQUIDs connected in series. The current through the SOT passes through a line which is inductively coupled to this array. This current induces a change in the flux through them, and as they are phase-coherent (if properly cooled down) this change in flux changes their critical current and accordingly the voltage on them. The SSAA has an additional FLL to keep the working point of the SQUIDs at their most sensitive (and most linear) region. Effectively, the

2.3 Scanning SQUID Microscopy

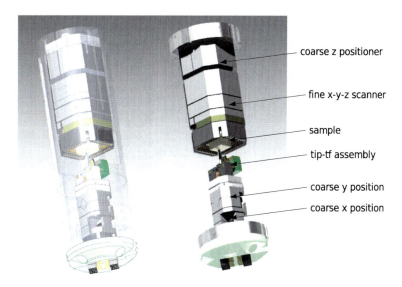

Fig. 2.5 Schematics of the microscope assembly. *Left* the microscope with the outer shell (made transparent for clarity); *Right* the microscope without outer shell. The lower part includes the coarse *x* and *y* positioners and the tip holder while the upper part includes the *z* positioner, the *xyz* scanners and the sample holder

SSAA functions as a flux-to-voltage converter, and with the SOT's current inductively coupled to a croygenic, low-impedance current-to-voltage converter. A scheme of the measurement circuit is given in Fig. 2.6. We voltage bias the SOT using the small parallel resistor (R_b) and measure the current through the SOT using the SSAA. Currently, the room-temperature feedback box is bandwidth limited to 1 MHz, which means that we cannot measure signals at frequencies higher than 500 kHz. Moreover, the SSAA current noise density is $\approx$2.5 pA/Hz$^{1/2}$, which sets our system noise limit (see Sect. 3.1.1)

2.3.1.2 The Tuning Fork and Its Feedback Loop

One can monitor the height of the resonance peak and set the feedback threshold to, for example, 50% of the maximum. This method, also known as slope detection [13], is not fast enough, since quartz tuning-forks have relatively high Q-factors, especially at low temperatures (typically 100,000) [13, 14]. As explained in Refs. [13, 15], a phase-locked loop (PLL) increases the bandwidth and makes scanning probe microscopy with tuning-forks possible at reasonable scan speeds. A scheme of the entire measurement system with emphasis on the PLL block-diagram is shown in Fig. 2.7. As already mentioned in Sect. 1.1.3.2, interaction forces between the tip and the sample result in a frequency shift of the tuning-fork's resonance. The basic idea behind the PLL is to track this resonance in a closed loop circuit and compare

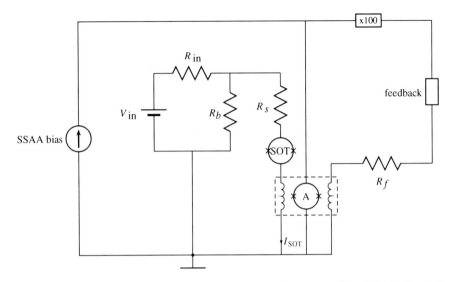

Fig. 2.6 The SOT circuit. Here "A" represents the SQUID series array amplifier (SSAA). The 5 kΩ series resistor, R_b and the SSAA are thermally coupled to the fridge's 1 K tube coil. The preamp (×100) and the feedback circuit are battery-operated room-temperature electronics. R_s is a parasitic series resistance to the SOT

its current location with the fundamental one to give the frequency shift, Δf. This frequency shift serves as the error function of the height, or z, feedback loop.

2.3.1.3 Coarse and Fine Motion of the Microscope

The coarse motion is performed by three attocube positioners (1 × ANPz101RES, 2 × ANPx50), one for each axis, based on the slip-stick [16] mechanism. The positioners themselves are driven by a saw-tooth pulse from a high-voltage driver (atto**cube** ANC-150). The z-axis positioner incorporates a resistive encoder, which allows for a resistive readout of the current displacement of the positioner. This resistor is calibrated so we can translate the value in Ohms to displacement in μm. Scanning is performed by an attocube integrated xyz scanner (ANSxyz100), with an $X - Y$ scan range of 30 × 30 μm² and a Z-scan range of 15 μm (all at low temperatures). We drive these piezoelectric scanners with high voltage from an SPM controller (either RHK SPM-100 or attocube ASC500).

The output of the PLL (in the form of a frequency shift, Δf) serves as the input of the z-height feedback loop (point no. 6 in Fig. 2.7). This second feedback loop (the first being the phased-locked one) controls the height of the tip above the sample by changing the voltage going to the z-scanner piezo.

2.3 Scanning SQUID Microscopy

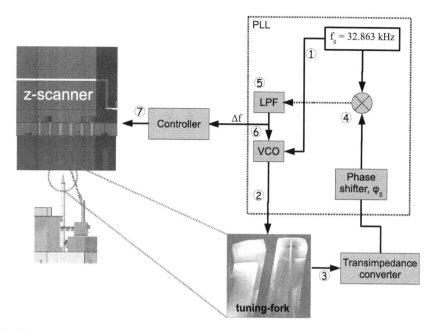

Fig. 2.7 A scheme of the height control loops. (*1*) The PLL sends an excitation voltage at the base resonant frequency, $\cos(f_0 t)$, to the VCO; (*2*) the VCO adds the base excitation phase together with any error corrections from the PLL feedback loop, and sends it, i.e. $\cos(f_0 t + \varphi_0)$, to the tuning-fork; (*3*) the current from the tuning-fork may experience a phase shift, φ, due to change in height and is then converted at room-temperature by a transimpedance converter; (*4*) the voltage from the converter is phase shifted by φ_0 and mixed with the base resonance frequency; (*5*) the mixed signal $\cos(2f_0 t + \varphi_0 + \varphi) + \cos(\varphi_0 + \varphi)$ is passed through a low-pass filter to produce the error signal, φ; (*6*) this error signal is fed back to the VCO and; (*7*) knowing the original resonance curve, e.g. Fig. 2.4, one can convert the phase shift into frequency shift, Δf, which serves as the feedback parameter for the height control loop

2.3.2 Approach Procedure

The microscope assembly's heart revolves, then, around the SOT and the TF, designed as a rigid structure holding the two, one against the other, each resting on either coarse motors or piezoelectric scanners. At room temperature the distance between the SOT/TF assembly and the sample is a few hundred of microns, and once they are cooled to low temperature, we initiate what is known as the approach procedure. This involves fully extending the z-piezo while simultaneously trying to close a feedback loop with the frequency shift as the set-point parameter (typically 100 mHz in our case). If the set-point is not reached, the z-piezo is fully retracted and the z-coarse motor moves by a few steps forward, repeating the entire procedure all over again. As with every feedback loop, there is the subtle interplay between the three (or two) parameters of the PID (PI) loop, known as "proportional", "integral" (and "derivative"). Usually, if one wants the loop to close in a reasonable time, an

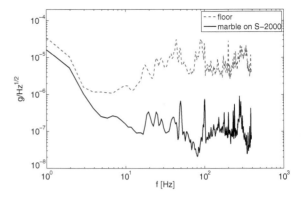

Fig. 2.8 Comparison of vibrations' level. The measurement was performed using a Wilcoxon 731A accelerometer, which measures acceleration in units of Earth's gravity field, g. The signal is then fed to a spectrum analyzer

overshoot above the desired value is obligatory. In the case of our SOT, however, an overshoot is highly dangerous, as the device itself, the SQUID, is located at the very edge of the tip, making it the first to be in potentially hazardous contact with the sample. This is also in practice what we had encountered in the first few approach sequences, all resulting in the tip crashing into the sample. In order to overcome this obstacle, we patterned our samples in a meander, or serpentine-like, shape. The original idea was to run current through the meander, so that the magnetic field resulting from this current (according to Biot-Savart law) would be observable tens of microns away from the sample. I will discuss this in more detail in Sect. 3.2. With the feedback loop closed we could then maintain a constant height above the sample and acquire images.

2.3.3 Vibration Isolation

If one wants to approach the surface of a sample within a few nanometers and maintain a constant height of a few nanometers above it, one needs to properly isolate the system from its surroundings, namely from vibrational, acoustical and electric noises. For structural vibration isolation, we placed the dewar on a massive marble block, weighing 920 kg. This block, in turn, was placed on four commercial isolators (Newport S-2000). These isolators attenuate vibrations from the environment at low frequencies, having already a 20 dB attenuation at 10 Hz. The disadvantage of these isolators is that they typically have a resonance, i.e, amplification, at around 1 Hz. To reduce acoustical noise, we first coated the entire dewar with an acoustipipe, thereby damping its self-resonating mode. In addition, we enclosed the entire upper part of the system (from the marble block and upwards) with a 45 dB attenuation factor of an acoustic enclosure. The combination of the isolators and the enclosure created a sufficiently quiet environment in which we were able to safely scan samples with a tip-sample separation of a few nanometers (Fig. 2.8).

2.4 Fabrication of Samples   27

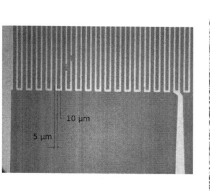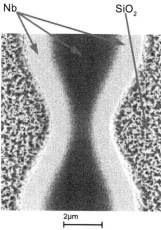

Fig. 2.9 *Left* an optical image of the Al serpentine deposited on a Si/SiO$_2$ substrate. The *arrows* indicate the direction of current running through it, in a serpentine trajectory; *Right* a SEM image of a Nb serpentine. It is very similar in shape and pattern, except for a constriction (shown) which repeats every 100 μm

2.4 Fabrication of Samples

The samples in this work were all prepared by our group in the clean-room facilities offered by our department. The mask itself is rather simple, having a serpentine structure going from one end to another. The main challenge, however, is to successfully fabricate such a structure with a minimum amount of defects so that current can flow unhindered from one side to the other. With a sample size of approximately 5 × 5 mm and a pitch of 15 μm (the width of the film is 5 μm and each two neighboring films are 10 μm apart), this amounts to a continuous line of the order of one meter!

The aluminum samples were deposited on a Si/SiO$_2$ substrate by e-gun evaporation in a background pressure of 1×10^{-6} Torr, and the photo resist (Shipley Microposit 1805) was removed using the lift-off technique. An optical image of the aluminum serpentine is given in Fig. 2.9. The niobium samples were deposited on a Si/SiO$_2$ substrate in the same vacuum chamber, with a substrate temperature of 300°C during deposition [17] and a rate of 0.3 nm/s. Then the Nb was capped with 3 nm of gold and the photolithography was performed using the previously mentioned 1805 photoresist. The gold layer was removed using argon ion-milling and finally the Nb was removed using reactive ion etching with SF$_6$ plasma. The complete recipe was taken from Ref. [18]. The mask used was slightly different, having a constriction (shown in Fig. 2.9) repeating itself approximately every 100 μm.

References

1. Yoo, M. J., Fulton, T. A., Hess, H. F., Willett, R. L., Dunkleberger, L. N., Chichester, R. J., Pfeiffer, L. N., and West, K. W. Scanning single-electron transistor microscopy: Imaging individual charges. *Science* **276**, 579 (1997).
2. Eytan, G., Yayon, Y., Bar-Joseph, I., and Rappaport, M. L. A storage dewar near-field scanning optical microscope. *Ultramicroscopy* **83**, 25 (2000).
3. Grober, R. D., Acimovic, J., Schuck, J., Hessman, D., Kindlemann, P. J., Hespanha, J., Morse, A. S., Karrai, K., Tiemann, I., and Manus, S. Fundamental limits to force detection using quartz tuning forks. *Rev. Sci. Instrum.* **71**, 2776 (2000).
4. Shelimov, K. B., Davydov, D. N., and Moskovits, M. Dynamics of a piezoelectric tuning fork/optical fiber assembly in a near-field scanning optical microscope. *Rev. Sci. Instrum.* **71**, 437 (2000).
5. Karrai, K. Lecture notes on shear and friction force detection with quartz tuning forks. (2000).
6. Cady, W. The piezo-electric resonator. *Proc. Inst. Rad. Eng.* **10**, 83 (1922).
7. Dye, D. W. The piezo-electric quartz resonator and its equivalent electrical circuit. *Proc. Phys. Soc. London* **38**, 399 (1925).
8. Zeltzer, G., Randel, J. C., Gupta, A. K., Bashir, R., Song, S.-H., and Manoharan, H. C. Scanning optical homodyne detection of high-frequency picoscale resonances in cantilever and tuning fork sensors. *Appl. Phys. Lett.* **91**, 173124 (2007).
9. Karrai, K. and Grober, R. D. Piezoelectric tip-sample distance control for near field optical microscopes. *Appl. Phys. Lett.* **66**, 1842 (1995).
10. Lupascu, A., Verwijs, C. J. M., Schouten, R. N., Harmans, C. J. P. M., and Mooij, J. E. Nondestructive readout for a superconducting flux qubit. *Phys. Rev. Lett.* **93**, 177006 (2004).
11. Kirtley, J. R. and Wikswo, J. P. Scanning SQUID Microscopy. *Annual Review of Materials Science* **29**, 117 (1999).
12. Welty, R. and Martinis, J. A series array of DC SQUIDs. *IEEE Trans. Magn.* **27**, 2924 (1991).
13. Albrecht, T. R., Grütter, P., Horne, D., and Rugar, D. Frequency modulation detection using high-Q cantilevers for enhanced force microscope sensitivity. *J. Appl. Phys.* **69**, 668 (1991).
14. Dürig, U., Steinauer, H. R., and Blanc, N. Dynamic force microscopy by means of the phase-controlled oscillator method. *J. Appl. Phys.* **82**, 3641 (1997).
15. Gildemeister, A. E., Ihn, T., Barengo, C., Studerus, P., and Ensslin, K. Construction of a dilution refrigerator cooled scanning force microscope. *Rev. Sci. Instrum.* **78**, 013704 (2007).
16. Renner, C., Niedermann, P., Kent, A. D., and Fischer, O. A vertical piezoelectric inertial slider. *Rev. Sci. Instrum.* **61**, 965 (1990).
17. Rairden, J. and Neugebauer, C. Critical temperature of niobium and tantalum films. *Proceedings of the IEEE* **52**, 1234 (1964).
18. Lichtenberger, A., Lea, D., and Lloyd, F. Investigation of etching techniques for superconductive Nb/Al-Al$_2$O$_3$/Nb fabrication processes. *IEEE Trans. Appl. Supercond.* **3**, 2191 (1993).

Chapter 3
Results

This work summarizes the efforts put into developing the first scanning SQUID microscope based on the SOT. As such, most of the results reported in this chapter are either of a preliminary characterization nature or involve extensive calibrations and testing of the system. In Sect. 3.1 I describe in detail the characterization of the aluminum SOT and to some minor extent that of SOTs made of other metals. Then, in Sect. 3.2 I present the data gathered while testing the system at its development stage.

3.1 SQUID-on-Tip Characterization

3.1.1 Aluminum SQUIDs

We invented a simple self-aligned fabrication method of a nanoSQUID on the apex of a sharp tip. This nanoSQUID is the smallest reported in literature to this day, with a loop diameter as small as ≈100 nm, as can be seen in Fig. 3.1. After its preliminary invention [1], we proceeded and characterized such an aluminum-based SOT at magnetic fields as high as 0.6 T (our aluminum becomes normal beyond this field) and up to currents which are two times as large as the critical current, I_c. Furthermore, the SOT was assembled in a prototype scanning SQUID microscope and was scanned above a meander-shaped aluminum thin film while running current through this meander, giving rise to a magnetic field profile due to this current and measured by the SOT several microns above it. These results were published in Ref. [2], attached below.

Self-Aligned Nanoscale SQUID on a Tip

Amit Finkler,*,[†] Yehonathan Segev,[†] Yuri Myasoedov,[†] Michael L. Rappaport,[†] Lior Ne'eman,[†] Denis Vasyukov,[†] Eli Zeldov,[†] Martin E. Huber,[‡] Jens Martin,[§] and Amir Yacoby[§]

[†]*Department of Condensed Matter Physics, Weizmann Institute of Science, Rehovot 76100, Israel, [‡]Departments of Physics and Electrical Engineering, University of Colorado, Denver, Colorado 80217, and [§]Department of Physics, Harvard University, Cambridge, Massachusetts 02138*

ABSTRACT A nanometer-sized superconducting quantum interference device (nanoSQUID) is fabricated on the apex of a sharp quartz tip and integrated into a scanning SQUID microscope. A simple self-aligned fabrication method results in nanoSQUIDs with diameters down to 100 nm with no lithographic processing. An aluminum nanoSQUID with an effective area of 0.034 μm^2 displays flux sensitivity of 1.8×10^{-6} $\Phi_0/Hz^{1/2}$ and operates in fields as high as 0.6 T. With projected spin sensitivity of 65 $\mu_B/Hz^{1/2}$ and high bandwidth, the SQUID on a tip is a highly promising probe for nanoscale magnetic imaging and spectroscopy.

KEYWORDS SQUID, microscopy, spin, detection, scanning

Imaging magnetic fields on a nanoscale is a major challenge in nanotechnology, physics, chemistry, and biology. One of the milestones in this endeavor will be the achievement of sensitivity sufficient for detection of the magnetic moment of a single electron.[1] There are three main technological challenges: fabrication of a sensor with a high magnetic flux sensitivity, reduction of the size of the sensor, and the ability to scan the sensor nanometers above the sample. Superconducting quantum interference devices (SQUIDs) have the highest magnetic field sensitivity, but their loop diameter is usually many micrometers. Much effort has been devoted recently to the development of nanoSQUIDs,[2−8] which have shown very promising flux sensitivity.[2−8] Most of these devices, however, are based on planar technology using lithographic or focused ion beam (FIB) patterning methods.[3,9,5,4,7,6,10,8,11] The large in-plane size of the devices precludes bringing the SQUID loop into sufficiently close proximity to the sample (due to alignment issues) to scan it with optimal sensitivity. Recently, a terraced SQUID susceptometer was developed that is based on a multilayered lithographic process combined with FIB etching. This device includes a 600 nm pickup loop which can be scanned 300 nm above the sample surface.[12] Here we present a simple method for the self-aligned fabrication of a dc nanoSQUID on a tip with effective diameter as small as 100 nm that can be scanned just a few nanometers above the sample.

We have fabricated several SQUID-on-tip (SOT) devices of various sizes. A quartz tube of 1 mm outside diameter is pulled to a sharp tip with apex diameter that can be controllably varied between 100 and 400 nm. The fabrication of the SOT consists of three "self-aligned" steps of thermal evaporation of Al, as shown schematically in Figure 1a. In the first step, 25 nm of Al is deposited on the tip tilted at an angle of −100° with respect to the line to the source. Then the tip is rotated to an angle of 100°, followed by a second deposition of 25 nm. As a result, two leads on opposite sides of the quartz tube are formed, as shown in Figure 1b. In the last step 17 nm of Al are evaporated at an angle of 0°, coating the apex ring of the tip. The two areas where the leads contact the ring form "strong" superconducting regions, whereas the two parts of the ring in the gap between the leads, indicated by arrows in Figure 1c, constitute two weak links, thus forming the SQUID. The resulting nanoSQUID requires no lithographic processing, its size is controlled by a conventional pulling procedure of a quartz tube, and it is located at the apex of a sharp tip that is ideal for scanning probe microscopy.

The studies were carried out at 300 mK, well below the critical temperature $T_c \approx 1.6$ K of granular thin films of aluminum in our deposition system. Instead of the commonly used current bias, the SOT was operated in a voltage bias mode, as shown schematically in the inset to Figure 2. We use a low bias resistance R_b of about 2 Ω and the SOT current, I_{SOT}, is measured using a SQUID series array amplifier (SSAA) working in a feedback mode.[5,13,14] R_s is a parasitic series resistance.

The resulting $I-V$ characteristics display several interesting features, as shown in Figure 2. First, the advantage of our SOT and the voltage bias setup is that there is no hysteresis, which avoids the need for complicated pulsed measurements.[15] Second, we observe a large negative differential resistance over a wide range of biases. This behavior is consistent with the Aslamazov−Larkin model of a single Josephson junction[16] if the voltage bias circuit of Figure 2 is taken into account. Third, small SQUIDs often have a small modulation of the critical current with field.[4,6]

* To whom correspondence should be addressed, amit.finkler@weizmann.ac.il.
Received for review: 01/5/2010
Published on Web: 02/04/2010

3.1 SQUID-on-Tip Characterization

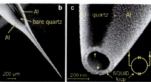

FIGURE 1. (a) Schematic description of three self-aligned deposition steps for fabrication of SOT on a hollow quartz tube pulled to a sharp tip (not to scale). In the first two steps, aluminum is evaporated onto opposite sides of the tube forming two superconducting leads that are visible as bright regions separated by a bare quartz gap of darker color in the SEM image (b). In a third evaporation step, Al is evaporated onto the apex ring that forms the nanoSQUID loop shown in the SEM image (c). The two regions of the ring between the leads, marked by the arrows in (c), form weak links acting as two Josephson junctions in the SQUID loop. The schematic electrical circuit of the SQUID is shown in the inset of (c).

Our SOT, in contrast, shows very pronounced $I_c(H)$ modulation as seen in Figure 2. Finally, the $I-V$ characteristics show fine structure at high biases, e.g., the 25 mT curve in Figure 2, which apparently results from resonances, the exact nature of which requires further investigation.

Figure 3a shows $I_{SOT}(V_{in}, H)$ plots displaying very pronounced quantum interference patterns with a period of 60.8 mT, corresponding to an effective SQUID diameter of 208 nm. The modulation of the critical current is large, $I_c^{max}/I_c^{min} = 1.67$, and in addition a large asymmetry between negative and positive biases is observed. Due to this asymmetry, the interference patterns at negative and positive bias are almost out of phase. From a theoretical fit[17] to $I_c(H)$, shown in Figure 3a by the dashed curves, we extract the following parameters: the critical currents of the two junctions $(1-\alpha)I_0 = 0.8$ μA and $(1+\alpha)I_0 = 2.4$ μA, where $I_0 = 1.6$ μA, the asymmetry parameter $\alpha = 0.5$, and $\beta = 2LI_0/\Phi_0 = 0.85$, where L is the loop inductance and $\Phi_0 = h/2e$ is the flux quantum. The asymmetric interference patterns therefore arise from the difference in the critical currents of the two junctions. This asymmetry is in fact very advantageous for scanning probe applications since high sensitivity can be attained essentially at any field by an appropriate choice of the SOT bias polarity and voltage.

The almost optimal $\beta = 0.85$ of the SOT implies a large inductance $L = 549$ pH. For comparison, the calculated geometrical inductance of our loop is $L_g = \mu_0 R(\log(8R/r) - 2) = 0.26$ pH, which is more than 3 orders of magnitude smaller. Here $R = 104$ nm is the loop radius and $r = 15$ nm is the radius of the loop wire. Our device is therefore governed by the kinetic inductance[18] of the loop, $L_k = 2\pi\mu_0\lambda_L^2 R/a$, due to its small dimensions. Here $a = 510$ nm^2 is the estimated cross sectional area of the loop, resulting in penetration depth $\lambda_L = 0.58$ μm. This λ_L is much larger than the bulk value for Al but is quite plausible for very thin granular Al films.[19,20]

Usually SQUIDs are operational only at very low fields. Remarkably, the SOT can operate over a very wide range of fields without a significant reduction in sensitivity. Parts b and c of Figure 3 show substantial quantum oscillations at fields as high as 0.5 T, which provides a unique advantage for investigation of various systems. This special property of SOT apparently arises from the fact that all the device

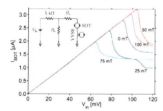

FIGURE 2. $I-V$ characteristics of the SOT at 300 mK at different applied fields. The inset shows the schematic measurement circuit. The SOT is voltage biased using a small bias resistor R_b, and the current I_{SOT} is measured using a SQUID series array amplifier (SSAA) with a feedback loop.

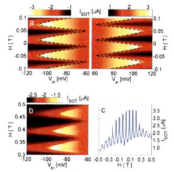

FIGURE 3. Quantum interference patterns of the SOT current $I_{SOT}(V_{in}, H)$ at 300 mK at positive and negative voltage bias. The patterns are asymmetric both in field and in bias and are almost out of phase for the two bias polarities. The dashed line shows a theoretical fit taking into account the difference in critical currents of the two weak links. (b) Quantum interference patterns at high fields up to 0.5 T. (c) Current oscillations $I_{SOT}(H)$ at a constant bias $V_{in} = 103.5$ mV over a wide field range.

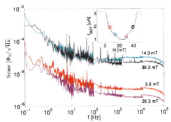

FIGURE 4. Spectral density of the flux noise of the SOT at 300 mK and different applied fields. The inset shows $I_{SOT}(H)$ at a constant bias $V_{in} = 100$ mV with the fields, indicated by colored circles, for which the noise spectra are presented. The lowest white noise level is $1.8 \times 10^{-6} \Phi_0/\text{Hz}^{1/2}$. The mismatch at 10^4 Hz is an instrumental artifact.

dimensions are very small and the thin superconducting leads along the quartz tube are aligned parallel to the applied field.

Figure 4 shows the spectral density of the flux noise of the SOT at various applied fields. Above a few tens of hertz the low frequency 1/f-like noise changes into white noise on the level of 3×10^{-5} to $1.8 \times 10^{-6} \Phi_0/\text{Hz}^{1/2}$ over a wide range of fields, which translates into a field sensitivity of 1.1×10^{-7} T. Our flux sensitivity is comparable to that of state of the art SQUIDs,[21] yet the area of the SOT loop is only 0.034 μm^2, which is the smallest reported to date.[4,9] The small size of SOT is highly advantageous for spin detection since spin sensitivity in units of $\mu_B/\text{Hz}^{1/2}$ is given by

$$S_n = \Phi_n \frac{R}{r_e}\left(1 + \frac{h^2}{R^2}\right)^{3/2}$$

where R is the radius of the loop, h is the height of the loop above the spin dipole, $r_e = 2.82 \times 10^{-15}$ m, and Φ_n is the flux sensitivity in $\Phi_0/\text{Hz}^{1/2}$.[12,22] For $h < R$ we obtain spin sensitivity of 65 $\mu_B/\text{Hz}^{1/2}$ for spins located in the center of the loop with an on-axis magnetic moment. In the SOT geometry, however, enhanced sensitivity could be achieved by imaging the spins near the perimeter of the loop.[23] In this case the sensitivity is mainly determined by the width of the weak link of about 30 nm rather than the diameter of the loop of 208 nm, leading to an estimated sensitivity of about 33 $\mu_B/\text{Hz}^{1/2}$. Such sensitivity should allow imaging of the spin state of a single molecule,[24] for example Mn_{12} acetate ($m = 20 \mu_B$), with integration of a few seconds. Moreover, our smallest operating SOT has an effective diameter of 130 nm. Assuming that it has the same flux noise as the larger SOT, the estimated spin sensitivity could be increased to below 20 $\mu_B/\text{Hz}^{1/2}$.

We have integrated the SOT into a scanning probe microscope operating at 300 mK in which the tip is glued

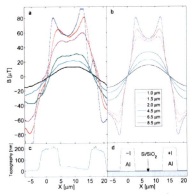

FIGURE 5. (a) Scanning SOT microscope measurement of a superconducting serpentine. Shown are the magnetic field profiles at various heights above the serpentine carrying a 2 mA current at 510 Hz. (b) Theoretically calculated field profiles at comparable indicated heights. (c) Topography across two strips of the serpentine based on feedback from a quartz tuning fork operating in constant height mode. (d) Schematic cross section of the serpentine showing the direction of the current in each strip.

to one tine of a quartz tuning fork. The frequency shift or the reduction in amplitude of the resonance peak of the tuning fork is used as a feedback mechanism for tip proximity to the sample surface.[25,26] This method provides the possibility of simultaneous imaging of sample topography and of the local magnetic field. The amplitude of the tip oscillation is typically less than 1 nm and, hence, does not degrade the spatial resolution of the magnetic imaging. As a test sample we used a 200 nm thick film of Al patterned into a serpentine structure. Figure 5c shows a topographical scan across two adjacent strips of the seprpentine using the tip and the tuning fork. A transport current of 2 mA at 510 Hz was applied to the sample, and the self-induced magnetic field was measured by the SOT at various heights above the surface. The results are shown in Figure 5a. The data are in good agreement with the theoretically calculated field profiles shown in Figure 5b. A field as low as 1 μT is readily measurable, which allows detection of the seprpentine signal from a distance of 10 μm above the surface.

In summary, we have developed a simple method for fabrication of sensitive nanoSQUIDs on the apexes of sharp tips and have incorporated them into a scanning SQUID microscope. A nanoSQUID with effective area of 0.034 μm^2 operated at fields as high as 0.6 T and displayed flux sensitivity of $1.8 \times 10^{-6} \Phi_0/\text{Hz}^{1/2}$, which translates into on-axis spin sensitivity of 65 $\mu_B/\text{Hz}^{1/2}$. By optimization of the SOT parameters, a further reduction in the noise can be expected, which, combined with a smaller loop diameter,

3.1 SQUID-on-Tip Characterization

could lead to a significant improvement in spin sensitivity. Compared to other SQUID technologies, the ability to image magnetic fields just a few nanometers above the sample surface renders the SQUID on tip a highly promising tool for study of quantum magnetic phenomena on a nanoscale.

Acknowledgment. We thank D. E. Prober, M. R. Beasley, I. M. Babich, and G. P. Mikitik for fruitful discussions. This work was supported by the European Research Council (ERC) Advanced Grant, by the Israel Science Foundation (ISF) Bikura Grant, and by the Minerva Foundation.

REFERENCES AND NOTES

(1) Rugar, D.; Budakian, R.; Mamin, H. J.; Chui, B. W. *Nature* **2004**, *430*, 329.
(2) Kirtley, J. R. *Supercond. Sci. Technol.* **2009**, *22*, No. 064008.
(3) Huber, M. E.; Koshnick, N. C.; Bluhm, H.; Archuleta, L. J.; Azua, T.; Björnsson, P. G.; Gardner, B. W.; Halloran, S. T.; Lucero, E. A.; Moler, K. A. *Rev. Sci. Instrum.* **2008**, *79*, No. 053704.
(4) Troeman, A. G. P.; Derking, H.; Borger, B.; Pleikies, J.; Veldhuis, D.; Hilgenkamp, H. *Nano Lett.* **2007**, *7*, 2152.
(5) Hao, L.; Macfarlane, J. C.; Gallop, J. C.; Cox, D.; Beyer, J.; Drung, D.; Schurig, T. *Appl. Phys. Lett.* **2008**, *92*, 192507.
(6) Lam, S. K. H. *Supercond. Sci. Technol.* **2006**, *19*, 963.
(7) Foley, C. P.; Hilgenkamp, H. *Supercond. Sci. Technol.* **2009**, *22*, No. 064001.
(8) Cleuziou, J.-P.; Wernsdorfer, W.; Bouchiat, V.; Ondarcuhu, T.; Monthioux, M. *Nat. Nanotechnol.* **2006**, *1*, 53.
(9) Granata, C.; Esposito, E.; Vettoliere, A.; Petti, L.; Russo, M. *Nanotechnology* **2008**, *19*, 275501.
(10) Wu, C. H.; Chou, Y. T.; Kuo, W. C.; Chen, J. H.; Wang, L. M.; Chen, J. C.; Chen, K. L.; Sou, U. C.; Yang, H. C.; Jeng, J. T. *Nanotechnology* **2008**, *19*, 315304.
(11) Girit, C.; Bouchiat, V.; Naaman, O.; Zhang, Y.; Crommie, M. F.; Zettl, A.; Siddiqi, I. *Nano Lett.* **2009**, *9*, 198.
(12) Koshnick, N. C.; Huber, M. E.; Bert, J. A.; Hicks, C. W.; Large, J.; Edwards, H.; Moler, K. A. *Appl. Phys. Lett.* **2008**, *93*, 243101.
(13) Welty, R.; Martinis, J. *IEEE Trans. Magn.* **1991**, *27*, 2924.
(14) Huber, M.; Neil, P.; Benson, R.; Burns, D.; Corey, A.; Flynn, C.; Kitaygorodskaya, Y.; Massihzadeh, O.; Martinis, J.; Hilton, G. *IEEE Trans. Appl. Supercond.* **2001**, *11*, 4048.
(15) Hasselbach, K.; Veauvy, C.; Mailly, D. *Physica C* **2000**, *332*, 140.
(16) Aslamazov, L. G.; Larkin, A. I. *JETP Lett.* **1969**, *9*, 87.
(17) Tesche, C. D.; Clarke, J. *J. Low Temp. Phys.* **1977**, *29*, 301.
(18) *Handbook of superconducting materials*; Cardwell, D. A., Ginley, D. S., Eds.; Institute of Physics Publishing: Bristol, U.K., and Philadelphia, PA, 2002.
(19) Cohen, R. W.; Abeles, B. *Phys. Rev.* **1968**, *168*, 444.
(20) Gershenson, M.; McLean, W. L. *J. Low Temp. Phys.* **1982**, *47*, 123.
(21) Kleiner, R.; Koelle, D.; Ludwig, F.; Clarke, J. *Proc. IEEE* **2004**, *92*, 1534.
(22) Ketchen, M.; Awschalom, D.; Gallagher, W.; Kleinsasser, A.; Sandstrom, R.; Rozen, J.; Bumble, B. *IEEE Trans. Magn.* **1989**, *25*, 1212.
(23) Tilbrook, D. L. *Supercond. Sci. Technol.* **2009**, *22*, No. 064003.
(24) Lam, S. K. H.; Yang, W.; Wiogo, H. T. R.; Foley, C. P. *Nanotechnology* **2008**, *19*, 285303.
(25) Atia, W. A.; Davis, C. C. *Appl. Phys. Lett.* **1997**, *70*, 405.
(26) Karrai, K.; Grober, R. D. *Appl. Phys. Lett.* **1995**, *66*, 1842.

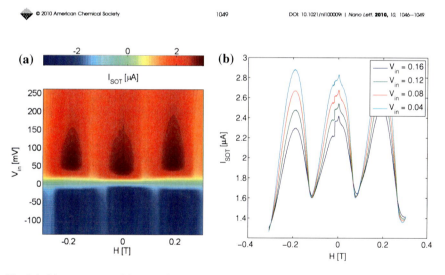

Fig. 3.1 Measurements of the world's smallest SQUID to date. **a** Mapping of the SOT current as a function of voltage bias and magnetic field. The period corresponds to a loop of diameter of 112 nm; **b** Several rows from (**a**), showing the I–V curves at different fields

Negative Differential Resistance The current–voltage characteristics in Fig. 2 of Ref. [2] exhibit a negative differential resistance over a wide range of voltage biases. The origin of this apparent decrease in resistance is the fast Josephson oscillations (in the GHz range), whose average we measure. As the current through the junction

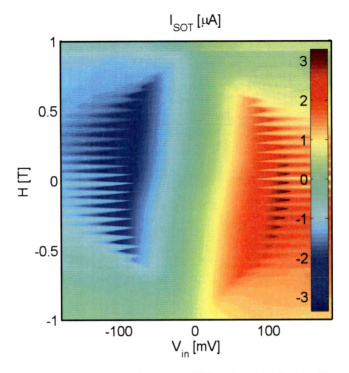

Fig. 3.2 A measurement of the current through the SOT as a function of voltage bias and field

reaches the critical current, a voltage develops on it, and the phase starts oscillating. The current we measure with the SSAA, therefore, is composed of two components, which are the maximal supercurrent, $I_s = I_c \sin\theta$, and the normal current, $I_n = \frac{1}{R}\frac{\hbar}{2e}\frac{\partial\theta}{\partial t}$. The time-development of the phase has been derived by Aslamazov and Larkin [3]. With the structure of the measurement circuit taken into account (see Appendix A) we see that the time-averaged current does in fact show a decrease at the critical current and only then an increase.

Wide Field Range Another illustrative example of the magnetic fields at which the SOT can work is given is Fig. 3.2. Quantum interference patterns are observed at fields as high as 0.7 T. We attribute this to both the granular nature of the aluminum film and its thickness.

Asymmetry As one can easily note from Fig. 3 in Ref. [2], there is a large asymmetry in both voltage bias and magnetic field. There are typically two major contributors to asymmetry—the difference between the critical current of the two Josephson junctions, I_1, I_2, and the difference in the inductance of the two arms comprising the SQUID loop, L_1, L_2. We define, accordingly, the asymmetry parameters, $I_2/I_1 = (1+\alpha_I)/(1-\alpha_I)$ and $L_2/L_1 = (1+\alpha_L)/(1-\alpha_L)$. Since the sum of the two currents is the critical current of the device, $I_1 + I_2 = I_C$, then, for example, an asymmetry of $\alpha_I = 0.2$ would imply a 50% difference between the two critical currents. A

3.1 SQUID-on-Tip Characterization

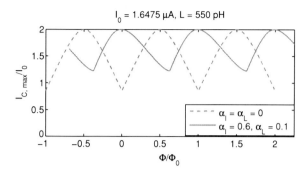

Fig. 3.3 Comparison between critical current-field curves of an ideal SQUID (no asymmetry) and one in which there is a large critical current asymmetry α_I, and a small inductance asymmetry, α_L

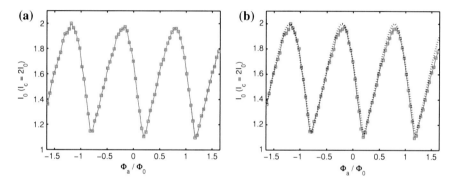

Fig. 3.4 **a** The critical current of the SOT as a function of magnetic field for positive voltage bias, taken from Ref. [2] and normalized both axes. The main feature of the asymmetry visible here is the shift from $\Phi_a = 0$ of the maximal critical current. Note also the different slopes of each period, depending on whether they are the first half or the second half; **b** This asymmetry can be modeled by taking into account asymmetries in the critical currents of the two junctions and the inductances of the two arms of the SQUID loop. The fitting parameters, using Ref. [4] notation, are $\alpha = 0.5$, $\beta = 0.85$ and $\eta = 0$

graph comparing two $I_c(\Phi)$ curves with different asymmetry parameters is plotted in Fig. 3.3. We try to fit our results with the theory. These results is plotted in Fig. 3.4a. The most prominent features are: (a) the shift from $\Phi_a = 0$ of the maximal critical current and; (b) the difference in slopes in one period depending on whether it is the first half of the period, e.g. from $\Phi_a/\Phi_0 = -0.8$ to $\Phi_a/\Phi_0 = -0.25$, or the second half, $\Phi_a/\Phi_0 = -0.25$ to $\Phi_a/\Phi_0 = 0.2$. The fit, using Ref. [4], is superimposed on the plot and is shown in Fig. 3.4b.

We tried taking this simulation one step further and take into account other components in the circuit, namely inductances and mutual inductances. The complete circuit analysis is described in Appendix B.

Table 3.1 Comparison of the different sources of noise (calculated)

Source	Noise (A^2/Hz)
Johnson	1.6×10^{-25}
Shot	3.2×10^{-25}
Quantum	1.3×10^{-25}

Sources of Noise Typically, four sources of noise are associated with fluctuations in dc-SQUIDs [4, 5]. We write them in the form of the spectral density of current noise, having the units of A^2/Hz:

1. Johnson noise $4k_BT/R$ in the normal state for $h\nu \lesssim k_BT$

$$S_I(\nu) = (2h\nu/R)\coth(h\nu/2k_BT) \simeq 4k_BT/R$$

   This noise is the result of thermal fluctuations in the normal part of the Josephson junction. R is the normal resistance of the junction and k_B is the Boltzmann constant.

2. Shot noise $S_I(\nu) = 2eI_0$. Becomes dominant above some voltage $V > k_BT/e$. According to Likharev [5], in weak links the condition to move from thermal noise to shot noise is $L \lesssim (v_F l T_\varepsilon)^{1/2}$, where v_F is the Fermi velocity, l is the mean free path and T_ε is the energetic relaxation time in the normal state.

3. Quantum noise when $h\nu > k_BT$, with $\nu = 2eV/h$. Then in that limit $S_I(\nu) = 2h\nu/R$, i.e. independent of temperature and in fact representing the zero-point fluctuations of an ensemble of harmonic oscillators with random phases [6].

4. $1/f$ or flicker noise. It is typically associated with magnetic flux fluctuations and critical current fluctuations. The origin of this noise in low-T_c superconductors is still unresolved (see Ref. [7] for more details).

For $R = 100\,\Omega$, $I_0 = 1\,\mu A$, $T = 300\,mK$ and $\nu = 10\,GHz$ (only then the condition of $h\nu > k_BT$ is satisfied) we can compare between the different sources of noises in Table 3.1.

Our system's noise floor was one order of magnitude larger (6.25×10^{-24} A^2/Hz), so there is definitely room for improvement. As one can see, quantum noise becomes relevant only at very low temperatures and very high frequencies.

Relaxation-Oscillation SQUID One feature of the SOT that had not been addressed in Ref. [2] is some high-frequency oscillations observed in the MHz range around the critical current. An example of such measurements in Pb-based SOTs is shown in Fig. 3.5. The tip shown in the figure was 241 nm in diameter, which corresponds to the period of 435 Oe. Together with Martin E. Huber from the University of Colorado, we hypothesized that this is in fact the manifestation of a relaxation-oscillation SQUID [8, 9]. It is a direct result of having the SQUID connected in series with the input coil of the SSAA. In short, a relaxation oscillation SQUID is a hysteretic SQUID connected to its voltage bias through an inductor and a resistor. If the bias current, I_b, is larger than the critical current I_c, relaxation oscillations occur, provided that $I_bR < V_g$, the gap voltage. The oscillations are the periodic increase

3.1 SQUID-on-Tip Characterization

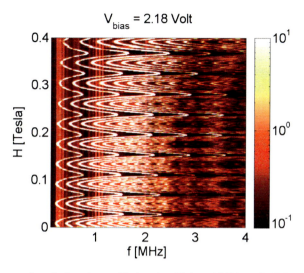

Fig. 3.5 Demonstration of relaxation-oscillations in a Pb-based SOT. A 10 μH inductor was intentionally connected in series with the SOT in order to create controlled conditions for the relaxation oscillations to appear in. The color code is a spectral density of voltage noise in units of $\mu V/Hz^{1/2}$. The *left-most* periodic pattern is the base oscillation ($f = 125$ kHz) and the ones to its *right* are its harmonics, which result from the voltage state switching nature of the oscillations. The period in magnetic field corresponds to the tip diameter, $d = 241$ nm

and decrease of the current through the SQUID due to switches in its voltage state (superconducting, normal) with a time constant $\approx L/R \times f\left(R, I_b, V_g, I_c(\Phi)\right)$. Due to this voltage state switching nature, we observe not only the base frequency but also all of its harmonics. Normally this time constant is very small and should not matter, as SQUIDs' inductance is typically small, on the order of pH. In our case, however, the SSAA's input coil's inductance is significant, approximately 0.25 μH, shifting the relevant frequencies to the MHz range. Using the circuit analysis of Ref. [8], the oscillation frequency is given by:

$$f = \frac{R_s}{L} \left\{ \log\left(\frac{I}{I - I_c}\right) + \log\left(1 + \frac{I_c}{V_g/R_s - I}\right) \right\}^{-1}.$$

Here, L is the inductance of our cryogenic Series SQUID Array Amplifier [10], R_s is a shunt resistance typically a few Ω, I_c is the critical current of the SQUID and V_g is the superconducting gap voltage, in our case approximately 250 μV. Using the numbers for our SQUIDs we indeed get a frequency of a few MHz, increasing with the biasing current, I. This, however, was only a qualitative test, and more experiments are needed to confirm this phenomenon. The advantage of these oscillations is that the signal itself is rather big, typically a few tens of mV [11], and changes in magnetic flux are converted by the relaxation-oscillator into changes in the frequency of oscillations, which should be easily detected by any frequency-demodulation technique or a commercial counter.

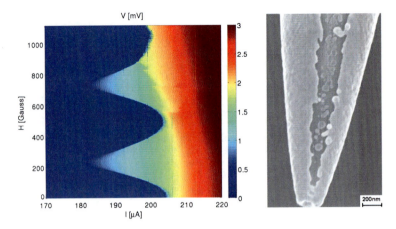

Fig. 3.6 *Left* Quantum interference patterns in a Pb-SOT. The SOT is current biased and the voltage drop on it is measured for different magnetic field values; *Right* A SEM image of a lead SOT

3.1.2 Pb, Nb and Sn SQUIDs

Apart from the aluminum SOTs, we have also tried to develop SQUID-on-tip sensors made from other superconductors, namely lead ($T_c = 7.2$ K), tin ($T_c = 3.7$ K) and niobium ($T_c = 9.2$ K). With thin ($d = 20$ nm) films of soft metals, such as lead and tin, one needs to cool the substrate before depositing lead or tin. Otherwise, the percolation threshold is too high (roughly 50 nm) to get a continuous film [12]. On the other hand, refractory metals such as niobium are highly sensitive to residual oxygen in the vacuum chamber during deposition, so that in order to deposit a superconducting niobium film in high vacuum one needs to heat the substrate to temperatures as high as 650–900°C [13, 14]. Alternatively, one can use an ultra high vacuum chamber and then there is less of a necessity for the heating of the substrate. Of the three metals, lead proved to be the simplest to develop. In the exclusive scope of this thesis we managed to successfully fabricate the **first** lead-based SQUID-on-tips. A typical measurement is given in Fig. 3.6. The continuation of this work has since been carried out by Drs. Denis Vasyukov and Yonathan Anahory, achieving nowadays a process of making Pb-based SOTs as small as 100 nm in diameter and a spin sensitivity lower than 20 $\mu_B/\sqrt{Hz}$.

3.2 Imaging

Having a reliable magnetic sensor, the next natural step was to test it with samples exhibiting relatively well-known physical properties, e.g. the Meissner effect in type-I superconductors. The first choice was a thin film of aluminum, patterned into a serpentine, followed by single crystals of NbSe$_2$ and thin films of niobium. Aluminum thin films, deposited on Si/SiO$_2$ substrates, are rather simple to fabri-

3.2 Imaging   39

Fig. 3.7 An image of the NbSe$_2$ crystal milled with a focused ion beam (FIB) to a pattern of a serpentine. The milled line width is approximately 5 μm

cate (see Sect. 2.4), and can be fabricated using a thermal evaporation technique in a background pressure of a few 10^{-6} Torr.

3.2.1 Magnetic Signals from Serpentines: Calibration Samples

As already described in Sect. 2.3.2, our first calibration sample was an aluminum serpentine, 200 nm thick, with a line width of 5 μm and a period of 15 μm (see Fig. 2.9). We ran a current of 2 mA through the serpentine and scanned the SOT over it. Already at a distance of 10 μm we were able to sense the magnetic field resulting from the current. In Appendix C we provide an algorithm for the calculation of the magnetic profile of a superconducting serpentine structure, which we compared to the profiles measured in Fig. 5 of Ref. [2].

Aluminum, however, is a type-I superconductor and therefore one cannot observe vortices in thick films of such materials. To that end, our next choice of a sample was a cleaved NbSe$_2$ crystal, milled with a focused ion beam (FIB) to a meander shape (see Fig. 3.7). The process of milling the meander pattern on such a crystal is lengthy (scale of several hours or even tens of hours). At some point one of the tips crashed, rendered the sample unusable and forced us to consider a simpler type-II superconducting material for the patterning of serpentines.

Finally, we fabricated high-quality Nb serpentines. With the installation of the vibration/isolation components, we succeeded in repeatedly acquiring images of the niobium sample. The tip-sample approach procedure was based on both the tuning-fork frequency shift for the last few nanometers but also on the magnetic signal from the niobium sample due to the Meissner effect: as one applies magnetic field on a

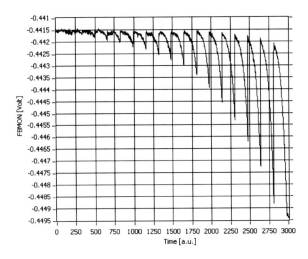

Fig. 3.8 "Live approach". The z-piezo element, on which the sample is mounted, slowly extends from zero to 15 μm, which is represented in this figure by the slow increase in magnetic signal (the sensitivity is 1 mV/Gauss). If there is no feedback signal from the tuning-fork, the z-piezo element is quickly retracted. This is the sharp drop in magnetic signal. Then the entire stage moves by a few coarse steps and the entire procedure repeats until a feedback signal is obtained. The size of a coarse step between two consecutive approaches is uncalibrated and lies between 100 nm and 1 μm

superconductor, Meissner currents are induced in it so as to negate the effect of the magnetic field on the superconductor and prevent flux lines from entering it. This results in a zero effective magnetic field just above the superconductor and *higher* than the applied external field outside of the edge of the strip. This way, as the tip approaches the sample, the effective magnetic field it feels increases if it is near the edge of a superconducting strip (see Fig. 2.9) and decreases to zero if it is directly above the center of it.

As described in Sect. 2.3.2, we set our approach sequence threshold to some value, above which the sequence stops and waits for the user's input. A "real-time" example of such an approach procedure is given in Fig. 3.8. It shows the amplitude of the signal measured by the SOT as a function of time. The amplitude increases in each iteration as the coarse motor makes a few more steps each time. Usually at this point, assuming the threshold values are set correctly, the tip is already only a few μm away from the sample. This is when we start acquiring images to see the magnetic signal from a large $30 \times 30\,\mu m^2$ area. Although there is no contact or closed feedback loop yet, images can still be acquired. Since one SOT may differ from another in its magnetic field sensitivity map, we usually perform a large field sweep each time we cool down a new SOT. This involves ramping the magnet to fields as high as ± 0.1 T. Consequently, magnetic flux in the form of vortices enters the sample and remains trapped there even after we lower the field back to zero. Therefore, images taken after such a field sweep will exhibit the trapped flux while

3.2 Imaging

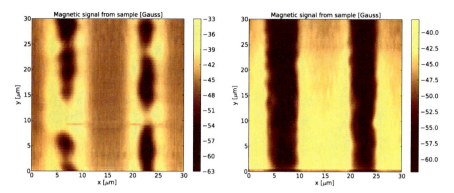

Fig. 3.9 Comparison between images with and without flux penetration. Both measurements were taken at $T = 0.3$ K and an applied field $H = -40$ Oe. The tip-sample separation was a few μm. *Left* The magnetic signal from a niobium serpentine with flux penetration. Maximum applied field before measurement: 1000 Oe. Bright regions are ones in which flux had fully penetrated. *Right* after cooling the sample in a magnetic field of -60 Oe

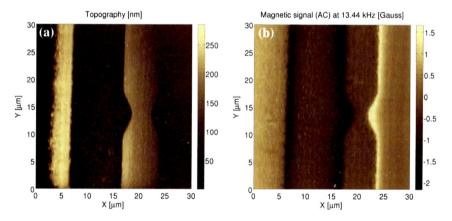

Fig. 3.10 **a** A topographic measurement of the Nb serpentine showing a double-edged funnel in the central part of the strip; **b** A self-field measurement of the same serpentine with a current of 3 mA at 13.44 kHz run through it

images taken after cooling at a specific magnetic field should have a vortex density proportional to the magnetic field, with a vortex–vortex separation of $\approx \sqrt{\Phi_0/B}$. This dramatic difference is shown in Fig. 3.9. In this figure, we first show the magnetic profile of the serpentine after sweeping the field between -1000 and 1000 Oe (left) immediately after the sequence from Fig. 3.8 terminated. As mentioned above, this usually means that the tip is a few μm above the sample. Then we heated the sample above its superconducting critical temperature (9.2 K), field-cooled it at an applied field of -60 Oe and measured the magnetic profile at an applied field of -40 Oe. At this tip-sample separation, one cannot resolve single vortices in niobium.

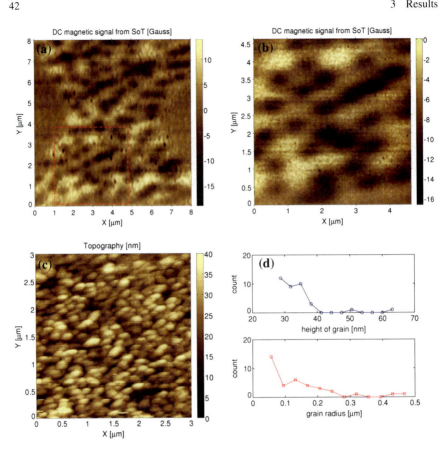

Fig. 3.11 **a** Magnetic image measured by the SOT a few nm above the funnel-shaped area in the serpentine taken after a field cooling in a magnetic field of −34 Gauss. The *dark spots* are vortices; **b** a measurement of the area marked by the *dashed red line* in the left image; **c** a topographic measurement of the Nb film, $3 \times 3\,\mu m^2$, in the same setup; **d** grain size analysis for (**c**). Most grains are between 120 and 300 nm in diameter and 30–40 nm in height

3.2.2 Vortices in a Niobium Serpentine

As a test sample we used a 200 nm thick niobium film, deposited using an e-gun while keeping the substrate at a temperature of 300 °C in a background pressure of 10^{-6} Torr and patterned as a meandering serpentine. With such a geometry one can drive a known current through the entire sample and measure its corresponding self-field while also be able to obtain the magnetic signal resulting from the Meissner effect and, of course, when close enough to the sample, observe vortices. We applied an AC current of 3 mA at a frequency of 13.44 kHz and measured the resulting self-field using the SOT concurrently with the topography measured from the tuning-fork's frequency shift. These two measurements are shown in Fig. 3.10. The self-field

3.2 Imaging

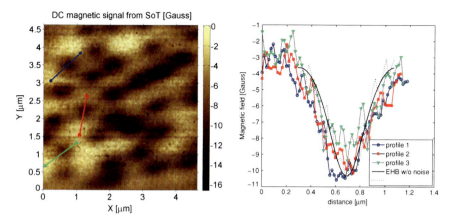

Fig. 3.12 Cross-sections of vortices in comparison to calculation. The best fit is for a magnetic penetration length, λ, of 400 nm and a coherence length, ξ, of 70 nm. This calculation takes into account the finite size of the tip, with a radius of 104 nm. The profile itself is calculated in an external field of 34 Oe and for a tip-sample separation of 20 nm. EHB are the initials of the late Prof. Dr. Ernst Helmut Brandt, who wrote the computer code for the calculation of these profiles at any field [16]. See Appendix D for a short explanation of how this profile is calculated, with emphasis on the incorporation of the finite tip size to the calculation

image agrees with the theoretical (Biot-Savart law) calculation of the magnetic field emanating from a current through a superconducting thin strip and closely follows the topography.

Figure 3.11a, b shows the DC magnetic signal a few nm above the double-edged funnel-shaped region in the serpentine shown in Fig. 3.10. The sample was field-cooled in an applied magnetic field of -34 ± 7 Oe. The vortex lattice is highly disordered, due to the strong pinning in Nb film at such a low temperature (see Ref. [15] and references therein). We can count approximately 34 individual vortices, which, for an image area of $4.6 \times 4.6\,\mu m^2$ gives a total field of 33 Gauss. The magnetic field modulation can be fit [16] to find the corresponding magnetic penetration length of the niobium film, which in this case turns out to be 400 nm. Figure 3.11c shows a topographic measurement of the same region, taken in our setup, showing the granular structure of the Nb film. The grain structure analysis is shown in Fig. 3.11d. Most grains are between 120 and 300 nm in diameter and 30–40 nm in height.

We took a few cross-sections of vortices from Fig. 3.11b and compared them to a simulated vortex profile calculated with $\lambda = 400$ nm, $\xi = 70$ nm, an applied field of 34 Oe and at a tip-sample separation of 10 nm. Figure 3.12 shows these cross-sections on the image from Fig. 3.11b and also superimposed for comparison with the calculated vortex profile.

References

1. Finkler, A. A SQUID on a tip: A tool to explore vortex matter in high-T_C superconductors. M.Sc. thesis, Weizmann Institute of Science, Rehovot, Israel, November (2006).
2. Finkler, A., Segev, Y., Myasoedov, Y., Rappaport, M. L., Ne'eman, L., Vasyukov, D., Zeldov, E., Huber, M. E., Martin, J., and Yacoby, A. Self-aligned nanoscale SQUID on a tip. *Nano Letters* **10**, 1046 (2010).
3. Aslamazov, L. G. and Larkin, A. I. Josephson effect in superconducting point contacts. *JETP Lett.* **9**, 87 (1969).
4. Tesche, C. D. and Clarke, J. DC SQUID: Noise and optimization. *J. Low Temp. Phys.* **29**, 301 (1977).
5. Likharev, K. K. Superconducting weak links. *Rev. Mod. Phys.* **51**, 101 (1979).
6. Koch, R. H., Van Harlingen, D. J., and Clarke, J. Quantum-noise theory for the resistively shunted Josephson junction. *Phys. Rev. Lett.* **45**, 2132 (1980).
7. Koch, R. H., DiVincenzo, D. P., and Clarke, J. Model for $1/f$ flux noise in SQUIDs and qubits. *Phys. Rev. Lett.* **98**, 267003 (2007).
8. Adelerhof, D. J., Nijstad, H., Flokstra, J., and Rogalla, H. (Double) relaxation oscillation SQUIDs with high flux-to-voltage transfer: Simulations and experiments. *J. Appl. Phys.* **76**, 3875 (1994).
9. Vernon, F. L. and Pedersen, R. J. Relaxation oscillations in Josephson junctions. *J. Appl. Phys.* **39**, 2661 (1968).
10. Welty, R. and Martinis, J. A series array of DC SQUIDs. *IEEE Trans. Magn.* **27**, 2924 (1991).
11. Mück, M. and Heiden, C. Simple DC-SQUID system based on a frequency modulated relaxation oscillator. *IEEE Trans. Magn.* **25**, 1151 (1989).
12. Wang, Z., Liu, Y., and Zhang, Z., editors. *Handbook of Nanophase and Nanostructured Materials*. Springer, (2002).
13. Neugebauer, C. A. and Ekvall, R. A. Vapor-deposited superconductive films of Nb, Ta, and V. *J. Appl. Phys.* **35**, 547 (1964).
14. Rairden, J. and Neugebauer, C. Critical temperature of niobium and tantalum films. *Proceedings of the IEEE* **52**, 1234 (1964).
15. Brandt, E. H. The flux-line lattice in superconductors. *Rep. Prog. Phys.* **58**, 1465 (1995).
16. Brandt, E. H. Ginzburg-Landau vortex lattice in superconductor films of finite thickness. *Phys. Rev. B* **71**, 014521 (2005).

Chapter 4
Discussion

In this work, I described the detailed characterization of our group's unique SQUID-on-tip. This all-aluminum device will hopefully be the precursor of more exciting variations of the sensor. It certainly has some very promising features for the field of magnetic scanning probe microscopy in terms of magnetic field sensitivity, sensor size and ease of manufacturing. While the projected spin sensitivity is currently "only" $65\,\mu_B/\sqrt{Hz}$, its followers, namely the all-lead SOT and others, show at least a ten-fold increase in sensitivity due to a much higher critical current. With those numbers in mind, one can easily think of studying molecular magnets [1, 2], Wigner crystals in carbon nanotubes [3, 4], and many other magnetic-rich phenomena on the microscopic scale.

The uncustomary geometry of the SOT together with coupling to a quartz tuning-fork allows for its assembly as a dual magnetic/topographic scanning probe microscope, making it the first SQUID acting as magnetic sensor in a scanning probe microscope with the sensor itself being only a few nanometers away from the surface of the sample. Since in many systems the magnetic signal decays strongly (distance cubed for magnetic dipoles, exponentially for a vortex lattice in superconductors, etc.), the sensor-sample separation is of great importance.

Indeed, this work culminated in successfully imaging magnetic phenomena in type-I and type-II superconductors, starting from verifying Biot-Savart's law in aluminum thin films, then confirming the Meissner effect in aluminum, $NbSe_2$ and niobium and finally imaging quantized flux, or vortices, in thin films of niobium. All of these measurements fill us with belief and confidence in the microscope's capabilities and future prospects.

I did not, however, manage to reach all of the goals defined in Sect. 1.3. Specifically, I did not succeed in observing vortex dynamics. The last sample I measured was a thin film of Nb, in which the pinning force at such a low temperature ($T/T_C = 0.03$) is very large, rendering vortex depinning virtually impossible within the limitation of maximal current that can be applied in our system due to heat dissipation vs. cooling power. There are several possibilities for materials in which the observation of vortex dynamics at 0.3 K is feasible, namely $NbSe_2$, FeSe, very thin films of aluminum, etc.

As our dear collaborator, Prof. Martin E. Huber from the University of Colorado, Denver, has said on numerous occasions in the past, this is just the first milestone in an exciting new project, which will surely lead to many important discoveries and hence, publications, in the future.

References

1. Sessoli, R., Gatteschi, D., Caneschi, A., and Novak, M. A. Magnetic bistability in a metal-ion cluster. Nature **365**, 141 (1993).
2. Friedman, J. R., Sarachik, M. P., Tejada, J., and Ziolo, R. Macroscopic measurement of resonant magnetization tunneling in high-spin molecules. Phys. Rev. Lett. **76**, 3830 (1996).
3. Deshpande, V. V. and Bockrath, M. The one-dimensional Wigner crystal in carbon nanotubes. Nature Phys. **4**, 314 (2008).
4. Secchi, A. and Rontani, M. Wigner molecules in carbon-nanotube quantum dots. Phys. Rev. B **82**, 035417 (2010).

Appendix A
Explanation of the Negative Differential Resistance

In this appendix we present a simplified analytical solution of the electric circuit used to measure the SOT. A more rigorous numerical solution of the SOT response is presented in Appendix B. We thank Grigorii Mikitik in realizing this analysis.

The model treats the SOT as a simplified Josephson junction, which is described by the critical current, I_c, phase difference, θ, between the two superconducting sides of the junction and a shunt resistor, R, which is the normal state resistance of the junction. As shown in Fig. A.1, the external circuit includes an input resistor, R_{in}, a parasitic resistance in series with the SOT, R_s, and the external voltage bias shunt resistor, R_b. The current through the SOT is denoted as I_{SOT}. The input resistor, R_{in}, is assumed to be much larger than all other resistors in the circuit. L is the inductance of the SSAA's input coil. Finally, I_{in} is the input current, which under the $R_{in} \gg R, R_b, R_s$ approximation, is just V_{in}/R_{in}.

First, we write the Kirchoff equations for the two cases, using the Josephson relation between the phase and voltage for the second case:

1. $I_{SOT} \leq I_c$

$$V_b = \frac{R_b \cdot R_s}{R_b + R_s} \cdot I_{in} \tag{A.1}$$

$$I_{SOT} = \frac{R_b}{R_b + R_s} \cdot I_{in} \simeq \frac{R_b}{R_b + R_s} \frac{V_{in}}{R_{in}} \tag{A.2}$$

2. $I_{SOT} > I_c$

$$V_b = (I_{in} - I_{SOT})R_b \tag{A.3}$$

$$V_b = I_{SOT}R_s + (I_{SOT} - I_c \cdot \sin\theta) \cdot R + \dot{I}_{SOT} \cdot L \tag{A.4}$$

$$(I_{SOT} - I_c \cdot \sin\theta)R = \frac{\hbar}{2e}\frac{\partial \theta}{\partial t} \tag{A.5}$$

Appendix A: Explanation of the Negative Differential Resistance

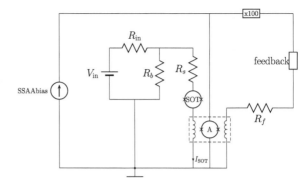

Fig. A.1 A scheme of the external circuit used to measure the SOT. The two inductors and the SQUID with an "A" inside comprise the SSAA. The boxes labeled "x100" and "feedback" are two custom-made electronic boxes, a preamp and a feedback box, respectively

In the limit of $L \to 0$, we use Eqs. A.1 and A.2, so that

$$(I_{\text{in}} - I_{\text{SOT}})R_b = I_{\text{SOT}} \cdot R_s + R(I_{\text{SOT}} - I_c \sin\theta) \Rightarrow I_{\text{SOT}} = \frac{I_{\text{in}} \cdot R_b + I_c R \sin\theta}{R_b + R_s + R}. \tag{A.6}$$

Using Eq. (A.6) in (A.5) gives us a differential equation for the phase as a function of the circuit parameters, $I_{\text{in}}, R_b, R_s, R$ and I_c:

$$\frac{\hbar}{2e}\dot\theta = \frac{R}{R_b + R_s + R} \cdot [I_{\text{in}} \cdot R_b - I_c \sin\theta(R_b + R_s)] \tag{A.7}$$

This equation can be written in a simpler form if we introduce the following notations:

$$\frac{\hbar}{2er}\dot\theta + I_c \sin\theta = \tilde{I}_{\text{SOT}} \tag{A.8}$$

with $r = R \cdot \frac{R_b + R_s}{R_b + R_s + R}$ and $\tilde{I}_{\text{SOT}} = I_{\text{in}} \cdot \frac{R_b}{R_b + R_s}$. This is in fact a differential equation of the form $a\dot y + b\sin(y) - c = 0$ with a known analytical solution

$$y(t) = 2\tan^{-1}\left\{\frac{b + \sqrt{c^2 - b^2}\tan\left[\frac{\sqrt{c^2 - b^2}}{2a}t\right]}{c}\right\},$$

so that we can write an exact expression for the phase and calculate its derivative:

$$\theta(t) = 2\arctan\left\{\sqrt{1 - \frac{I_c^2}{\tilde{I}_{\text{SOT}}^2}} \cdot \tan\left[e \cdot r \cdot t\frac{(\tilde{I}_{\text{SOT}}^2 - I_c^2)^{1/2}}{\hbar}\right] + \frac{I_c}{\tilde{I}_{\text{SOT}}}\right\} \tag{A.9}$$

$$\dot\theta \frac{\hbar}{2er} = \frac{\tilde{I}_{\text{SOT}}(\tilde{I}_{\text{SOT}}^2 - I_c^2)}{\tilde{I}_{\text{SOT}}^2 + I_c^2 \cos\omega t + I_c\sqrt{\tilde{I}_{\text{SOT}}^2 - I_c^2}\sin\omega t} \tag{A.10}$$

with $\omega = \frac{2er}{\hbar}(\tilde{I}_{\text{SOT}}^2 - I_c^2)^{1/2}$.

Appendix A: Explanation of the Negative Differential Resistance

Fig. A.2 $I_{SOT}(V_{in})$ characteristics of a SOT and a fit using $R = 90\,\Omega, R_b = 2.6\,\Omega, R_s = 1.25\,\Omega, I_c = 11.5\,\mu A$

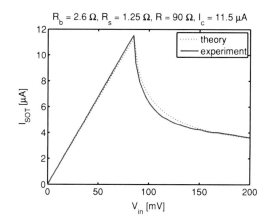

Using Eq. (A.10) in (A.8) we get

$$I_c \sin\theta = -\dot{\theta}\frac{\hbar}{2er} + \tilde{I}_{SOT} = \tilde{I}_{SOT} \cdot \frac{I_c^2(1+\cos\omega t) + I_c\sqrt{\tilde{I}_{SOT}^2 - I_c^2}\sin\omega t}{\tilde{I}_{SOT}^2 + I_c^2\cos\omega t + I_c\sqrt{\tilde{I}_{SOT}^2 - I_c^2}\sin\omega t}, \quad (A.11)$$

so that finally inserting back into Eq. (A.6)

$$\boxed{I_{SOT}(t) = I_{in} \cdot \frac{R_b}{R_b+R_s+R} + \frac{R}{R_b+R_s+R}\tilde{I}_{SOT}\left[1 - \frac{\tilde{I}_{SOT}^2 - I_c^2}{\tilde{I}_{SOT}^2 + I_c^2\cos\omega t + I_c\sqrt{\tilde{I}_{SOT}^2 - I_c^2}\sin\omega t}\right]}$$

$$(A.12)$$

Over one period, the average current through the SOT is

$$\boxed{\bar{I}_{SOT} = \frac{1}{2\pi}\int_0^{2\pi} d(\omega t) I_{SOT}(t) = \tilde{I}_{SOT} - \frac{R}{R_b+R_s+R}\sqrt{\tilde{I}_{SOT}^2 - I_c^2}} \quad (A.13)$$

To see that the current through the SOT decreases immediately after it reaches I_c, we can write $\tilde{I}_{SOT} = I_c + \epsilon$ for a small $\epsilon \ll I_c$:

$$\bar{I}_{SOT} = I_c + \epsilon - \frac{R}{R_b+R_s+R}\sqrt{2\epsilon I_c} \approx I_c\left(1 - \frac{R\sqrt{2}}{R_b+R_s+R}\sqrt{\frac{\epsilon}{I_c}}\right)$$

$\Rightarrow \bar{I}_{SOT}$ decreases with ϵ.

In our experimental setup, we measure I_{SOT} as a function of V_{in}. Therefore, if we want to summarize:

$$\bar{I}_{\text{SOT}}(V_{\text{in}}) = \begin{cases} \dfrac{V_{\text{in}}}{R_{in}} \dfrac{R_b}{R_b+R_s} - \dfrac{R}{R_b+R_s+R}\sqrt{\left[\dfrac{V_{\text{in}}}{R_{in}}\dfrac{R_b}{R_b+R_s}\right]^2 - I_c^2} & \dfrac{V_{\text{in}}}{R_{in}}\dfrac{R_b}{R_b+R_s} > I_c \\ \dfrac{V_{\text{in}}}{R_{in}}\dfrac{R_b}{R_b+R_s} & \dfrac{V_{\text{in}}}{R_{in}}\dfrac{R_b}{R_b+R_s} < I_c \end{cases}$$

(A.14)

In Fig. A.2 we plot $I_{\text{SOT}}(V_{\text{in}})$ of a tip and a fit according the derivation above, with $R = 90\,\Omega, R_b = 2.6\,\Omega, R_s = 1.25\,\Omega, I_c = 11.5\,\mu A$. The fit shows a reasonable match to the experiment. In this specific measurement the $I_{\text{SOT}}(V_{\text{in}})$ characteristics were not measured up to input voltages high enough in order to observe the eventual rise in current. This, of course, indeed occurs.

Appendix B
Full SOT Circuit Analysis

In this appendix we consider the SOT circuit in more detail, as presented in Fig. B.1. The model relies on each Josephson junction behaving according to the resistively- and capacitively-shunted junction (RCSJ) model [1, 2]. In this model, each junction has a critical current I_0, in parallel with its self-capacitance C and resistance R, so that the current through the entering these elements is $I = C\dot{V} + V/R + I_0 \sin\delta$.

We define the following currents using the Josephson relation between the current and the phase:

Current through...	Notation
C_L	$I_{CL} = C_L \dot{V}_{CL}$
C_R	$I_{CR} = C_R \dot{V}_{CR}$
Left junction	$I_{JL} = I_{0L} \sin\delta_L$
Right junction	$I_{JR} = I_{0R} \sin\delta_R$
L_L	I_{LL}
L_R	I_{LR}
Left weak link	I_{TL}
Right weak link	I_{TR}

We also define "flux quantum bar" (FQB), $\bar{\Phi}_0 \equiv \Phi_0/2\pi$, so that from the Josephson relation between the voltage and the phase we get

$$\boxed{\dot{\delta}_L = \frac{V_{CL}}{\bar{\Phi}_0}} \tag{B.1}$$

$$\boxed{\dot{\delta}_R = \frac{V_{CR}}{\bar{\Phi}_0}} \tag{B.2}$$

The current through the inductor and the voltage on it are, respectively,

A. Finkler, *Scanning SQUID Microscope for Studying Vortex Matter in Type-II Superconductors*, Springer Theses, DOI: 10.1007/978-3-642-29393-1,
© Springer-Verlag Berlin Heidelberg 2012

Fig. B.1 The SOT circuit in more detail, specifically the two Josephson junctions

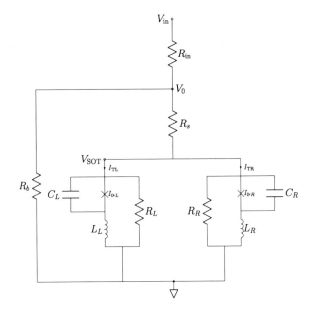

$$I_{LL} = I_{CL} + I_{JL} = C_L \dot{V}_{CL} + I_{0L} \sin \delta_L \tag{B.3}$$

$$V_{LL} = L_L \dot{I}_{LL} = L_L C_L \ddot{V}_{CL} + \dot{\delta}_L L_L I_{0L} \cos \delta_L. \tag{B.4}$$

Next, we want to write down an expression for the voltage on the junction (or capacitor) as a function of the current through the inductor and the phase. In order to do that we need to consider the following:

$$V_0 = \frac{V_{in} - R_{in}(I_{TL} + I_{TR})}{1 + R_{in}/R_b} \tag{B.5}$$

and

$$V_{SOT} = V_0 - R_s(I_{TL} + I_{TR}) = \frac{V_{in} - R_{in}(I_{TL} + I_{TR})}{1 + R_{in}/R_b} - R_s(I_{TL} + I_{TR}) \tag{B.6}$$

On one hand

$$I_{LL} = I_{TL} - \frac{V_{SOT}}{R_L} \tag{B.7}$$

and on the other hand

$$I_{LL} = C_L \dot{V}_{CL} + I_{0L} \sin \delta_L$$

so that

Appendix B: Full SOT Circuit Analysis

$$\boxed{\dot{V}_{CL} = \frac{1}{C_L}I_{LL} - \frac{I_{0L}}{C_L}\sin\delta_L} \quad (B.8)$$

$$\boxed{\dot{V}_{CR} = \frac{1}{C_R}I_{LR} - \frac{I_{0R}}{C_R}\sin\delta_R} \quad (B.9)$$

Finally, we want to write the equations for the time-derivative of the current through the inductor as a function of the voltage on the capacitor and the phase using $V_{SOT} = V_C + V_L$ and Eq. (B.6). In order to write that in a simple form we need first to write V_{SOT} as a function of I_{LL} and I_{LR} by inserting Eq. (B.7) into (B.6):

$$V_{SOT} = \frac{V_{in}}{1 + R_{in}/R_b} - \left(R_s + \frac{R_{in}}{1 + R_{in}/R_b}\right)\left[I_{LL} + I_{LR} + \left(\frac{1}{R_L} + \frac{1}{R_R}\right)V_{SOT}\right]$$

$$\Rightarrow V_{SOT} \underbrace{\left[1 + \left(R_s + \frac{R_{in}}{1 + R_{in}/R_b}\right)\left(\frac{1}{R_L} + \frac{1}{R_R}\right)\right]}_{p'}$$

$$= \underbrace{\frac{V_{in}}{1 + R_{in}/R_b}}_{q'} - \underbrace{\left(R_s + \frac{R_{in}}{1 + R_{in}/R_b}\right)}_{R'_{si}}(I_{LL} + I_{LR})$$

$$\Rightarrow V_{SOT} = \underbrace{\frac{q'}{p'}}_{1/p}V_{in} - \underbrace{\frac{R'_{si}}{p'}}_{R_{si}}(I_{LL} + I_{LR}) = \frac{V_{in}}{p} - R_{si}(I_{LL} + I_{LR}) \quad (B.10)$$

As we wrote earlier,

$$V_{SOT} = V_{CL} + V_{LL} = V_{CL} + L_L\dot{I}_{LL}, \quad (B.11)$$

so that

$$\dot{I}_{LL} = \frac{V_{SOT}}{L_L} - \frac{V_{CL}}{L_L}. \quad (B.12)$$

Inserting the simplified expression for V_{SOT} from Eq. (B.10) we get that

$$\boxed{\dot{I}_{LL} = \frac{1}{pL_L}V_{in} - \frac{R_{si}}{L_L}(I_{LL} + I_{LR}) - \frac{1}{L_L}V_{CL}} \quad (B.13)$$

$$\boxed{\dot{I}_{LR} = \frac{1}{pL_R}V_{in} - \frac{R_{si}}{L_R}(I_{LL} + I_{LR}) - \frac{1}{L_R}V_{CR}} \quad (B.14)$$

These six boxed Eqs. (B.1), (B.2), (B.8), (B.9), (B.13) and (B.14) comprise a set of six differential equations with six variables, δ_L δ_R V_{CL} V_{CR} I_{LL} and I_{LR}. As is standard in numerical analysis, and in exactly the same way as in Ref. [3], we switch to the following dimensionless units: voltage in units of I_0R, flux in units of Φ_0, current in units of I_0, capacitance in units of C_0, inductance in units of L_0, resistance in units of R and time, θ, in units of $\Phi_0/2\pi I_0 R$, so that

$$\frac{d}{dt} = \frac{2\pi I_0 R}{\Phi_0}\frac{d}{d\theta}.$$

The convention used is that the normalized form of a variable x will appear as $\tilde{x}$. We start with Eq. (B.1):

$$\dot{\delta}_L = \frac{d\delta_L}{dt} = \frac{2\pi I_0 R}{\Phi_0}\frac{d\tilde{\delta}_L}{d\theta} = \frac{I_0 R}{\Phi_0/2\pi}\tilde{V}_{CL},$$

so that

$$\boxed{\dot{\tilde{\delta}}_L = \tilde{V}_{CL}} \qquad (B.15)$$

$$\boxed{\dot{\tilde{\delta}}_R = \tilde{V}_{CR}} \qquad (B.16)$$

Next we look at Eq. (B.8):

$$\frac{dV_{CL}}{dt} = \frac{1}{C_L}I_{LL} - \frac{I_{0L}}{C_L}\sin\delta_L = \frac{I_0}{C_0\tilde{C}_L}\left(\tilde{I}_{LL} - \tilde{I}_{0L}\sin\delta_L\right),$$

so that

$$\boxed{\dot{\tilde{V}}_{CL} = \frac{d\tilde{V}_{CL}}{d\theta} = \frac{\Phi_0}{2\pi I_0 R^2 C_0}\frac{1}{\tilde{C}_L}\left(\tilde{I}_{LL} - \tilde{I}_{0L}\sin\delta_L\right)} \qquad (B.17)$$

$$\boxed{\dot{\tilde{V}}_{CR} = \frac{d\tilde{V}_{CR}}{d\theta} = \frac{\Phi_0}{2\pi I_0 R^2 C_0}\frac{1}{\tilde{C}_R}\left(\tilde{I}_{LR} - \tilde{I}_{0R}\sin\delta_R\right)} \qquad (B.18)$$

Last, we look at Eq. (B.13):

$$\dot{I}_{LL} = \frac{I_0}{\Phi_0/2\pi I_0 R}\dot{\tilde{I}}_{LL} = \frac{1}{p}\frac{I_0 R}{L_0}\frac{\tilde{V}_{in}}{\tilde{L}_L} - \frac{I_0 R}{L_0}\frac{\tilde{R}_{si}}{\tilde{L}_L}\left(\tilde{I}_{LL} + \tilde{I}_{LR}\right) - \frac{I_0 R}{L_0}\frac{\tilde{V}_{CL}}{\tilde{L}_L}.$$

Introducing $\beta_0 = 2L_0 I_0/\Phi_0$, we get

$$\boxed{\dot{\tilde{I}}_{LL} = \frac{1}{\pi\beta_0\tilde{L}_L}\left[\frac{1}{p}\tilde{V}_{in} - \tilde{R}_{si}\left(\tilde{I}_{LL} + \tilde{I}_{LR}\right) - \tilde{V}_{CL}\right]} \qquad (B.19)$$

$$\boxed{\dot{\tilde{I}}_{LR} = \frac{1}{\pi\beta_0\tilde{L}_R}\left[\frac{1}{p}\tilde{V}_{in} - \tilde{R}_{si}\left(\tilde{I}_{LL} + \tilde{I}_{LR}\right) - \tilde{V}_{CR}\right]} \qquad (B.20)$$

The relation between applied flux and the parameters is

$$\frac{\Phi_a}{\Phi_0} = \frac{1}{2\pi}(\delta_L - \delta_R) + \frac{1}{\Phi_0}(L_L I_{LL} - L_R I_{LR}). \qquad (B.21)$$

Appendix B: Full SOT Circuit Analysis

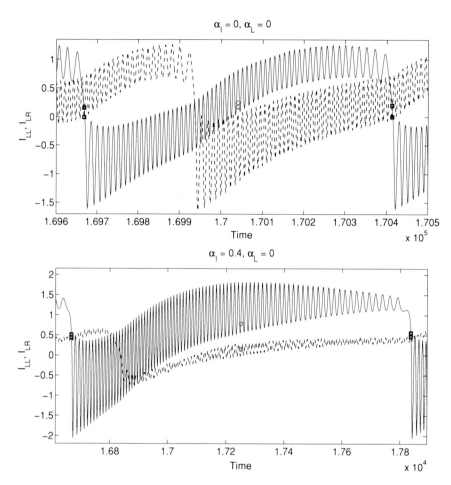

Fig. B.2 Current through *left* (I_{LL} in *solid*) and *right* (I_{LR} in *dashed*) inductors of each weak link for different parameters in the simulation. All calculations were performed for $\Phi/\Phi_0 = 0.4$. *Top* no asymmetry in any of the components (inductors, resistors, capacitors, critical currents); *Bottom* An asymmetry in the critical current ($I_{0L} = I_0(1 + \alpha_I)$ $I_{0R} = I_0(1 - \alpha_I)$); Each subfigure shows one typical period, corresponding to a Φ_0 flux slip through the SOT, as a function of time. The *rectangles* mark the beginning and the end of the period, identified by zero crossing. *Circles* mark the middle of the period

The initial conditions at DC are the following: The current in each junction does not pass through R or C, so that

$$I_{LL} = I_{JL} = I_{0L} \sin \delta_L = I_{TL}. \tag{B.22}$$

Therefore, Eq. (B.21) becomes

$$\frac{\Phi_a}{\Phi_0} = \frac{1}{2\pi}(\delta_L - \delta_R) + \frac{1}{\Phi_0}(L_L I_{0L} \sin \delta_L - L_R I_{0R} \sin \delta_R). \tag{B.23}$$

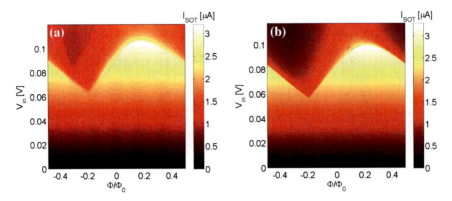

Fig. B.3 Comparison of normalized flux, Φ/Φ_0, voltage bias, V_{in} and current through SOT, I_{SOT}, characteristics; **a** Experimental data for one period in magnetic flux. With a period of 608 Gauss, this corresponds to a field span of ± 304 Gauss around zero; **b** Simulation based on Tesche and Clarke [7] according to the derivation given above. The asymmetry parameters used here were $\alpha_I = 0.5, \alpha_R = \alpha_C = \alpha_L = 0$, with $C = 0.1\,\text{pF}, L = 550$ pH and $R = 95\,\Omega$.

In our dimensionless units, Eq. (B.23) becomes

$$2\pi\Phi = \delta_L - \delta_R + \pi\beta_0(L_L I_{0L} \sin\delta_L - L_R I_{0R} \sin\delta_R). \tag{B.24}$$

where Φ is the dimensionless applied flux. From Eq. (B.24) we see that for a given applied flux and critical currents of both junctions, one can solve it, i.e. find for which values of δ_L and δ_R the equation is obeyed, and calculate the maximal critical current of the device.

For each value of flux, we then solve the six differential equations using an advanced Runge-Kutta method [4, 5] to get a matrix of solutions as a function of time. In principle this should be enough, since the solutions include the currents through each arm of the SOT, and we are interested in their sum. However, the currents are highly oscillatory in nature, and a simple averaging cannot give the correct average (DC) value which corresponds to the values measured in the experiment. A few examples of how these currents look like as a function of time for different asymmetry parameters is shown in Fig. B.2. Indeed the first observation is that the time dependence, i.e. before averaging, is non-trivial and changes dramatically when changing the asymmetry parameters. The second observation is that a simple averaging of this signal would not work. This calls for a more sophisticated way of averaging. It entails the identification of a "period" in the signal by looking at zero-crossings and then averaging over one period. Once we get the average (DC) values, we perform a linear interpolation of the arrays. The reason for that is due to the non-linear part of the array (above the critical current), the spacing between adjacent pixels is not even. For the DC values then, we compare results of this simulation with $\alpha_I = 0.5$, $\alpha_C = \alpha_R = \alpha_L = 0$ to data from Ref. [6]. This comparison is displayed in Fig. B.3.

Appendix C
Magnetic Field Profile of a Serpentine

In this appendix we show the steps of the calculation for the magnetic profile of a superconducting serpentine structure when current I is passed through it. The width of a strip in this calculation is a and distance between two near strip edges is b so that the period is $a + b$. A schematic drawing of the structure is shown in Fig. C.1.

The calculation is performed as follows:

1. Take a single strip and divide it into N segments. This defines the width of the discrete current element contributing to the magnetic field profile.
2. Create an array of x-coordinates, $[x]$, for the above single strip, having the same size, N.
3. Create the current density array, $[J(x)]$. Take into account the superconducting nature of the structure, i.e. current flows mostly on the edges of the superconductor [8]. Quantitatively, this amounts to multiplying the current elements' magnitude with a factor $1/\sqrt{(a/2)^2 - x^2(k)}$, where $x(k)$ goes from $-a/2$ to $a/2$. We see that indeed the biggest contribution comes from the edges, where this factor is largest.
4. Duplicate both $[x]$ and $[J(x)]$ to the left and to the right of the single strip, with the current density arrays multiplied by a minus sign to accommodate for the change of direction in the flow of current. Repeat this process several times (may be important for a narrow strip and short period, but is less important for a wide strip and large period). This duplication is actually an approximation, which assumes that there is no interaction between the strips.
5. Calculate the magnetic field contribution of each current element for a point $P(x_k, z)$, located a distance z from the surface of the serpentine and at location x_k along the x-axis. This calculation is performed using the Biot-Savart law, i.e.

$$\mathbf{B} = \int \frac{\mu_0}{4\pi} \frac{I d\mathbf{l} \times \hat{\mathbf{r}}}{|r^2|},$$

Appendix C: Magnetic Field Profile of a Serpentine

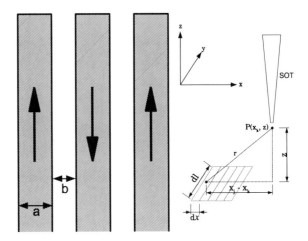

Fig. C.1 A schematic drawing of the serpentine structure. The superconducting strip has a width a and the structure itself a period of $a + b$. On the right we show the distances entering the current-element contribution to the magnetic field at a point $P(x_k, z)$.

where d*l* is a vector of length d*l* pointing along the current flow direction, r is the distance between the current element and the point P and μ_0 is the magnetic permeability of vacuum. For each point, $P(x_k, z)$, we need to add the contributions of all current-carrying elements along the x-axis *and* along the current flow direction, y. The latter is the usual Biot-Savart calculation for the magnetic field of a current-carrying wire, which gives $B = \mu_0/(2\pi r)$. For the former, we write the distance between each current-carrying element j and $P(x_k, z)$:

$$r = \sqrt{z^2 + (x_j - x_k)^2}.$$

Finally, it is important to remember that the SOT measures only flux perpendicular to it, or in our case, in the z-direction. Therefore, each contribution needs to be multiplied by its z-axis projection, or $\frac{(x_j - x_k)}{r}$. Thus, the magnetic field contribution to the signal measured by the SOT at a point $P(x_k, z)$ from a current carrying element at a point x_j is

$$d\mathbf{B}(x_k, x_j, z) = \frac{\mu_0}{2\pi} \frac{1}{\sqrt{z^2 + (x_j - x_k)^2}} \frac{(x_j - x_k)}{r}. \qquad (\text{C.1})$$

The field at $P(x_k, z)$ is therefore $\int d\mathbf{B}(x_k, x_j, z) dx_j$ where the limit is taken from one strip to the left of x_k's strip to one strip to its right.

Appendix D
Magnetic Field Profile of a Vortex as Seen by the SOT

We calculate the magnetic field of a Ginzburg-Landau vortex lattice in a superconductor having a coherence length, ξ, and magnetic penetration length, λ, for an applied external field H_{ext}. For this purpose we use a computer program written by Prof. Ernst Helmut Brandt [9]. In this program he uses Fourier series as trial functions for the Ginzburg-Landau function $|\psi(x,y)|^2$ and magnetic field $B(x,y)$ and minimizes the Ginzburg-Landau free energy with respect to a finite number of Fourier coefficients. The input parameters of this program are ξ, λ, H_{ext} and the film thickness, d, and the outputs are $\psi(x,y,z)$ and $B(x,y,z)$.

Our purpose is to accommodate for the finite radius of the SOT, R, into the actual magnetic field image we expect to measure. In image-processing jargon this means a convolution of the computed $B(x,y,z)$ with a kernel the size of the SOT's diameter. We denote the computed magnetic field for a specific height z above the sample as B_{kl}, which is an $N_x \times N_y$ matrix, with $k = 1\ldots N_x, k \in \mathbb{Z}$ and $l = 1\ldots N_y, l \in \mathbb{Z}$. For each element in this matrix we then compute the following double sum:

$$B_{kl} = \sum_{i=k-m_x}^{k+m_x} \sum_{j=l-m_y}^{m+m_y} B_{ij},$$

where $(x_i - x_k)^2 + (y_j - y_k)^2 < R^2$ and m_x and m_y are the (rounded) radii of the tip in pixels corresponding to the computed matrix, e.g. a 256×256 matrix of size $4 \times 4\,\mu m^2$ and a radius of 100 nm give $m_x = m_y = 7$ pixels.

References

1. McCumber, D. E. Effect of ac impedance on dc voltage-current characteristics of superconductor weak-link junctions. *J. Appl. Phys.* **39**, 3113 (1968).
2. Stewart, W. C. Current-voltage characteristics of Josephson junctions. *Appl. Phys. Lett.* **12**, 277 (1968).
3. Lichtenberger, A., Lea, D., and Lloyd, F. Investigation of etching techniques for superconductive Nb/Al-Al$_2$O$_3$/Nb fabrication processes. *IEEE Trans. Appl. Supercond.* **3**, 2191 (1993).
4. Bank, R., Coughran, W., Fichtner, W., Grosse, E., Rose, D., and Smith, R. Transient simulation of silicon devices and circuits. *IEEE Transactions on Computer-Aided Design of Integrated Circuits and Systems* **4**, 436 (1985).
5. Hosea, M., and Shampine, L. Analysis and implementation of TR-BDF2. *Appl. Num. Math.* **20**, 21 (1996).
6. Finkler, A., Segev, Y., Myasoedov, Y., Rappaport, M. L., Ne'eman, L., Vasyukov, D., Zeldov, E., Huber, M. E., Martin, J., and Yacoby, A. Self-aligned nanoscale SQUID on a tip. *Nano Letters* **10**, 1046 (2010).
7. Tesche, C. D., and Clarke, J. DC SQUID: Noise and optimization. *J. Low Temp. Phys.* **29**, 301 (1977).
8. Zeldov, E., Clem, J. R., McElfresh, M., and Darwin, M. Magnetization and transport currents in thin superconducting films. *Phys. Rev. B* **49**, 9802 (1994).
9. Brandt, E.H. Ginzburg-Landau vortex lattice in superconductor films of finite thickness. *Phys. Rev. B* **71**, 014521 (2005).

Index

A
Aluminum, 29
Asymmetry, 34

B
Bragg glass, 2
Broad-band noise, 13

D
Depinning force, 3
Dither piezo, 19

E
Effective spring constant, 12
Elastic channels, 6

F
Feedback Loop, 23
Field-cooled, 43
Fluxcreep, 3
Funnel, 43

I
Imaging, 38
Impedance mismatch, 22
Isolators, 26

J
Josephson equations, 7

L
Lead, 38
Lorentzian, 11
Low, 20

M
Magnetic profile, 41
Meander, 29
Mechanism, 24

N
Negative Differential Resistance, 47
Niobium, 38
Noise, 36

P
Phase-locked loop, 23
Piezoelectric scanners, 24
Pipette puller, 17

R
Relaxation oscillation, 36
Resistively- and capacitively-shunted junction, 51

S
Serpentine, 26, 27, 39, 42, 57
Shear-force microscopy, 12
Slip-stick, 24

S (*cont.*)
SQUID, 8
SQUID series array amplifier, 22

T
Tin, 38
Topography sensor, 18
Trans-impedance converter, 22
Tuning fork, 9

V
van der Waals, 11
Viscosity, 20
Vortex lattice, 59
Vortices, 1, 41, 42

W
Washboard frequency, 13

Printed by Publishers' Graphics LLC
MO20120615